THIS ISN'T FINE.

THIS ISN'T FINE.

The cultural shift we need
TO SAVE HUMANITY
from the dumpster fire we started

(a primer to the end of the world as we know it.)

taylor ahlstrom

foreword by nate hagens

—emdashery books—

For our children, their children, and all the ones after . . .
I sure hope this works.

ISBN: (Paperback) 979-8-9850333-6-6
ISBN: (Hardcover) 979-8-9850333-7-3
ISBN: (eBook) 979-8-9850333-8-0

Book interior & cover design by Taylor Ahlstrom.

"Question Hound" meme and cover artwork
licensed with permission from K.C. Green.

First printing edition 2024.

What's Cookin' Good Lookin'?
(it's the table of contents)

Foreword

by Nate Hagens

Our society shares socially acceptable stories about the future. But these 'memes' are increasingly disconnected from our biophysical reality. We misunderstand systems, and how the parts and processes of the global human economy interrelate. We misunderstand energy—the fact that all life, all commerce, any movement against gravity or inertia, or any generation of order, is an energetic process and is enabled or delimited by available net energy—i.e. energy availability sets our physical limits. We misunderstand ecology and how the entire human system is supported by a fragile (and fraying) interconnected web of life and its hidden but critical ecological functions. These misunderstandings have simple explanations, but massive implications for our future.

We live as part of a system, and how this system works is not taught in our schools—not really.

Your Dad might have told you where babies come from, but did he tell you where money comes from? (Most dads, including mine, don't know). Money is created out of thin air when commercial banks make loans. When they do this, the financial system gets bigger, but the world we live in stays the exact same size. The same amount of oil, copper, water, forests, and fish existed as minutes before. And yet we create millions upon millions of dollars every single day without ever thinking about what that money truly represents.

And every dollar that changes virtual hands around the globe represents not only the physical materials to create the thing, but the energy behind it . . . energy to imagine, to invent, to refine, to manufacture, to deliver, to run, to maintain, to repair, and to dispose of. There is not a single activity in our modern economy that doesn't require energy. But did you know 85 percent of that energy still comes from fossils in the forms of coal, gas, and oil? Did you know that one barrel of crude oil contains over 20,000 hours of your physical labor potential? For (still) less than $100?? And that the average American eats 2,500 kilocalories per day but we 'eat' over 200,000 kilocalories per day in fossil fuels? This stuff that powers our modern economies is non-renewable on human time scales. Everyone reading this page is alive at what one day may be referred to as The Carbon Pulse, where we are drawing down Earth's fossil energy battery millions of times faster than it trickle-charged by daily photosynthesis during our planet's deep past.

And as we endlessly grow this system and blindly burn through our finite stores of both energy and materials, the prices we pay fail to reflect the true value—or harm—of what we're doing. Since 1970 we have lost 70 percent of the populations of animals, birds, fish, and insects. Human sperm count is dropping 1-2 percent a year due to endocrine disrupting chemicals. Humans and our livestock outweigh wild animals by 50 times. Plastic created by our culture now outweighs wild animals. There is no escaping it: we need to entirely re-prioritize most of what we're doing.

I am a former college professor turned host of the podcast, *The Great Simplification*. For twenty years I've been putting together facts like this that comprise the systems synthesis of 'the human predicament'. Historically, humans have solved problems by adding complexity. Added complexity requires additional energy. As our fossil energy stores decline, what we do with technology is going to have to change. We will have to simplify our lives, our expectations, and our social interactions. This 'great simplification' will be one of the most momentous events of human history—and if you're reading this book, you will live through it.

This could be an exciting and upbeat time, the beginning of a true "age of sentience" for mankind. It could be humanity's finest hour . . . or it could go in other ways. The stakes couldn't be higher. The end of the world as we know it is *not* the end of the world. The coming decades will be a time of chaos, disruption, emergence, discovery, resilience, courage, purpose, and even joy and satisfaction. What matters is that we get started. And to do that, we first have to understand the situation—and to care about it. The book that follows by my friend Taylor Ahlstrom helps towards both of those goals. She has compiled an engaging and accessible summary of the real state in which we humans find ourselves, both as a culture and as a species, circa 2024 CE. But most of all, she's provided an inspiring and actionable path to a better future.

After reading, I invite you all to help us meet the future halfway. Our story is still being written.

The Dumpster Fire We Started

The Walking Worried

**Don't worry about the world coming to an end
today. It's already tomorrow in Australia.**

—Charles M. Schulz

I know I'm not alone when I say that I fell into a deep sort of depression during COVID. My husband and I were living in South Africa, and we were six weeks into what was supposed to be a two-week ban on alcohol and cigarettes . . . and leaving the house and walking the dog . . . and . . . everything. The booze had run out, I was surviving on black market cigarettes, and everything in the world felt hopeless. I watched as America told everybody to stay home but didn't give them any money to do it. I watched as half of America suggested we sacrifice our grandparents to save the economy. My work had dried up, and so I spent my days doomscrolling, just watching pain and suffering and poverty and hunger and race riots and death—so much death.

But when I think about it now, the discontent that had burrowed deep within me started long before COVID. And not to point the most obvious and oft-pointed finger, but it started sometime during the Trump administration. I have always cared about politics since I was old enough to realize that just because my parents were conservative

Christians, it didn't mean I was one as well. My years in college were spent in lively political debates that have lasted well into my ten-year marriage to a man with a master's in political science. Our courtship involved a lot of that same heated but friendly discussion and games of devil's advocate, and most of our marriage has continued in that fashion. Yes, we are liberal. No, I'm not really a Democrat, but what other choice do I have in a two-party system? Before you put down this book, let me assure you, it is most certainly not political. I believe both sides (as if there are only two sides) have a reason to be upset; both sides are feeling the effects of the things I'm talking about. *Everyone* is feeling the effects of the things I'm talking about. So please, bear with me.

No matter which team you root for, the years since Trump's election have been undoubtedly trying. They've been full of heartbreak as I longed to be able to do something but didn't know what the hell I could actually do. I watched depression rates skyrocket and opioids kill hundreds of thousands a year and a sharp uptick in suicides labeled as "deaths of despair" for an overly anxious, depressed, and generally fucked-up population. And all the while, wildfires are raging and "once-in-a-century" floods are happening yearly, and literal locust plagues of Biblical proportions are descending on East Africa, and polar vortexes, and police brutality, and Nazis making a comeback, and school shooting after school shooting after school shooting, and my heart just kept breaking again and again and again and again. I was overburdened by the weight of a million things I didn't believe could ever change. I watched the rise of authoritarianism around the world from France to China to Hungary to Italy—to America. I watched the invention of the term post-factualism and watched as people denied indisputable facts from reputable sources right before their eyes. I watched flat earthers attempt to argue that when the simple experiments they had devised to prove the flatness of the earth instead proved—once again—that it was round, it was the experiment that was at fault, and not the assumption.

And then I guess we sort of got used to Trump. We got used to how ridiculous it all was, every single day. We got used to the racism and the cronyism and the nepotism and the incompetence. We tried to impeach him when it got real out of hand, but that didn't work. And we got used

to that too. We got used to Russian interference and Chinese interference, and nobody really cared about those things in the first place,
anyway.

And then Ruth Bader Ginsberg died, and we lost the right to
choose, and the term Christofacism started popping up a lot. How far
would they go? How many more rights would they take away? Why do
they keep talking about pronouns when no one has fucking healthcare?
Oh right ... it's so we don't pay attention to the fact that no one has
healthcare. And that they're systematically defunding education. And
busting unions. And exploiting every last one of us. And then there was
"stop the steal" and lies about voter fraud and dozens and dozens of
meritless lawsuits, each one threatening to topple our fragile democracy.
And then there was the insurrection that almost worked.

There was a time when bipartisanship was championed in the
United States, and now it will see you straight to the gallows. There was
a time when politicians thought that they had to do what they said they
were going to do—until they all realized that, in fact, they do not. You
only have to say the thing, and you can do whatever you want. Because
the majority of people don't actually bother to find out how you voted
or who lined your pockets or who took away their healthcare or voted
against $25 insulin. The majority of people (if they watch news at all)
only watch one news source that already aligns with their beliefs, and we
have no laws surrounding media bias or reporting facts in our country,
so you can say and do whatever the hell you want and get away with it.
All you have to do is call the other people the enemy. All you have to
do is identify your tribe, and they'll vote for you no matter how many
secret abortions you paid for or hookers you slept with or what wars
you supported or didn't support. It's as true of the far right as much as
the far left, and now it feels as if those two sides have grown into an
unbridgeable chasm.

And believe me—if you're a conservative of any kind, I know you're
feeling things too. You're probably feeling lied to and cheated, and you're
seeing caravans of migrants swarming the border and lawlessness and
riots and the "moral decline of America." Everybody's upset. Everybody's
struggling. That's kind of the point.

And then Biden was elected, which literally no Democrat cared about other than that he wasn't Trump, and then Portland melted and Texas froze and California burned and Europe burned—and the Amazon was burning the whole fucking time—and in my heart it was all too much. I couldn't hold the weight of it. There was no answer. The extreme tribalism of our political system was irreparable. No one cared. No one was doing anything to fix any of it. We could all literally see the world burning, and we couldn't even agree on the color of the flames.

And if through all this there was maybe one thing that could have brought our nation together across party lines—in the same way it happened after 9/11—it was a global pandemic that affected every human being equally regardless of race or creed or nationality. But instead of uniting us, the big blond Cheeto decided to use it to split us further apart. If humanity wasn't even willing to wear a piece of cloth over their faces to save someone else's life, then surely, we as a species were doomed. The global response to COVID was a low fucking bar, and we utterly failed. At least the murder hornets never came.

And after years of feeling like the world was getting worse and worse—and then just when you thought it couldn't get worse, it did—I broke. All of these things jammed into lockdowns and COVID depression and relative unemployment led me to feel a way I had never felt before: I just stopped caring. No, it's not that. I stopped being *able* to care. My heart just couldn't break again for all the things I knew I couldn't fix. I stopped reading the massive bills being passed in Congress that were never going to help anyone anyway. I stopped engaging in political conversations with my husband, telling him, "I just can't care about that right now." I stopped attempting to engage in calm political debate with people on Brietbart or Fox News or my crazy uncle (that never really worked anyway). I stopped trying to change anybody's mind or make a difference. We moved to Spain in the middle of the pandemic, and I started trying to . . . enjoy life. I unsubscribed from the *Washington Post*'s "Daily 202" round-up. I unsubscribed from *The New Yorker* and *The Economist* and *The New York Times*. I stopped reading Fox News and Brietbart. I stopped reading anything, really. I just drank wine and explored a new city (after lockdown was lifted and we were allowed to leave our apartment, of course) and thought, man, life is so much better here in Spain. I mean, we're experiencing a massive, multi-year drought and a "once-in-a-century" blizzard, and oh shit, it's already over 100 degrees in May???

But at least we have healthcare. And cheap wine.

And then as soon as the pandemic was over (it's over, right?) there was historic inflation and stagflation and shrinkflation strangely aligned with record corporate profits, and still nobody seemed to do anything about that either. Memes were posted, people talked about how messed up it all was and is, but once again, nothing changed. The minimum wage is *still* $7.25. The rich got richer and the poor got poorer, and we shared some more memes about how inequality is higher than it was during the French Revolution or the Gilded Age or fucking feudalism . . . and that was it. Just a meme and a chuckle at the futility of it all, and we all went back to paying our bills (if we could) and just trying to survive.

And through this hazy, disconnected sort of malaise I had created, I watched the world as if I wasn't in it. I watched it like news of a war in a country on the other side of the planet (though I was barely following the one that was actually happening). I mused about the collapse of capitalism and if I might be here to see it. What would it look like? And I drank and danced and partied and had fun and pretended like not a single thing was wrong within my tiny world. When faced with the massive, unending, seemingly unfixable tragedies of the world, it felt like hedonism was a reasonable response.

Until the summer of 2022. In the summer of 2022, my husband introduced me to a podcast called *The Great Simplification*. We were on a six-week trip around Southeast Asia with lots of time on long bus rides, so we started listening. And suddenly, I was snapped out of my intentioned escapism. Suddenly, everything started to make sense. All the things wrong, everything collapsing and nobody caring, it wasn't politics or social media or tribalism—I mean, it is those things—but it is so much more.

It is a million things I didn't realize I didn't know. Things we take for granted as the only way the world has ever been. Things about how money works, how energy works, and how our economy works. It's the first cracks in the foundation of the wildly complex yet largely invisible systems we created to keep our oh-so-incredible society running oh-so-smoothly. It is paying the piper for the hundred years of exponential growth we've recklessly enjoyed. It is the beginnings of the end of the world as we know it, and there isn't a damn thing anyone can do about it. Well, there is. But no one is actually doing it. Nate Hagens, the host of that podcast, put into perspective what I had been unable to put my finger on this whole time. All of the problems we're facing are part of this one, great, gigantic problem that literally nobody is talking about. And when you finally start to understand how precariously everything is glued together, it's painfully easy to see how it's all about to fall apart.

As I listened to episode after episode of this arguably depressing podcast, rather than fall deeper into hopeless tragedy, I became more and more energized. I snapped out of my self-imposed detachment and, for the first time in so many years, felt like I had a purpose. Actually, for the first time in my life. Something bigger than me, something bigger than you. Something bigger than all of us because it's everything. It's literally everything on the line.

So, I started writing this book. I started writing, hoping beyond hope it could do the same for you. Hoping to show you what I could finally see and snap you out of the haze of the walking worried. Hoping to help you get to post-tragic. Because that's the secret sauce. If we wanna save the world, we have to see past all the doomsday prophecies and apocalyptic futures. Past the problems we've been trying to solve that are just downstream symptoms of the sick we've got. We have to understand that the answers are a whole lot simpler than they seem. And they're completely, unquestionably possible.

But the road to post-tragic drives right through the tragedy. And I hate to break it to you, but it's gonna get bleak. Some of you won't even finish this book. I guess I wouldn't blame you if you just put it down right now and decide to read a trashy romance novel or a formulaic spy thriller or just binge-watch some escapist TV. As hilarious as I may be at times, there's only so much comic relief can do to lighten the mood. But I promise you, if you don't put this book down, if you grin and bear it through all the messed up things that are even more messed up than everyone has been saying they are . . . if you manage to make it through all the reasons why all the solutions we think we have aren't going to fix anything at all . . . there *is* light at the end of the tunnel. And a lot of it. It's goddamn radiant when you remember how absolutely miraculous we humans can be.

This isn't a depressing book about the end of the world. Well, it is a little bit. But it's a book about hope. It's about who we are and how we got here and what we've forgotten. It's about all the answers that are right in front of us. It's about the fact that whether you're feeling overwhelmed by the tragedy, whether you've given in or given up, whether you're in a hedonism phase or a nihilism phase or you never gave a shit about any of this in the first place . . . there's something—there are a lot of things—we can do to turn this ship around.

If we want to put out the fire, we gotta figure out what's burning. And if we want to build something better, we have to know which pieces were better off getting burnt. So, stick with me, and we might just be able to save the world.

The Incredible Everything Machine

It's not just that this coming half-century will bear so little resemblance to the last one. It's that the last half-century bears so little resemblance to almost any other time and place in human history itself that's confusing us.

—Jamie Wheal
Recapture the Rapture

Most of us probably don't think much about this world we live in. We go to work and school, we watch our TV shows, we order Uber Eats when we're feeling lazy, and we scroll TikTok or Instagram or whatever so we never get bored. We stare at a medium screen all day before we turn on the bigger screen at night so we can look at our small screen while the big one plays in the background. This all seems pretty normal to us. But our entire lives—and most of our parents' lives—have existed during the pinnacle of human existence. During the pinnacle of ease and convenience and same-day delivery and 150 different flavors of ranch dressing. It wasn't always like this. In fact, it almost never was.

I was born in 1984, which means I am part of the "Oregon Trail" generation of elder millennials who at least remember being teenagers without the internet. The first website I remember visiting was in the year 1997, and I distinctly recall typing in www.leonardodicaprio.com. I used to print out MapQuest directions to drive to my friends' houses, and before that, we had to scribble down directions on a scrap of paper like "turn left at the big tree." In high school, I had a beeper.

But even before the dawn of the internet and GPS and an app for everything, the last seventy years of human history have been absolutely fucking bonkers. The Industrial Revolution was huge for industry and travel and manufacturing, sure. It made farming easier and machines faster and increased productivity and connected goods and services to ever more distant places. But it was also the massive shift toward globalization that was made possible by America in the aftermath of World War II that created this absolute paradise of convenience we're living in now.

Let's take a moment to bask in the marvel of the American grocery store. Every time I travel home to the US, I am the most excited to go to the grocery store. Nowhere else in the world is there such a vibrant, incredible, colorful, mind-boggling display of consumer choice. According to Michael Ruhlman, author of *Grocery: The Buying and Selling of Food in America*,[1] the average grocery store in America has more than 40,000 individual products on its shelves. FORTY-FUCKING-THOUSAND! In the 1800s? It was just two hundred. And even as recently as the 1990s, there were a paltry 7,000 items for you to choose from. In just thirty years, we've increased our selection by nearly six-fold.

In 1989, then Russian President Boris Yeltsin famously visited an American grocery store and said if the Russian people (who were waiting in food lines at the time) saw this cornucopia . . . there would be a revolution.

In case you've never traveled outside of the US, I can assure you, the rest of the world, even the developed world, is not like this. You want salad dressing? They've got two. You want a frozen pizza? Maybe two brands each with three different flavors. In America, DiGiorno alone now sells seven different types of CRUSTS. And each of those crusts comes with myriad options for toppings—pepperoni, supreme, spinach and garlic, four-cheese, FIVE-cheese ... whatever. DiGiorno sells a whopping *seventy-eight* varieties of frozen pizza. And they're just a single brand in a massive aisle dedicated almost entirely to frozen pizzas. This shit is INSANE.

To put how wild all this is into perspective, let's rewind a bit. Back before we discovered oil for our lamps (mostly from whales we had to go out and murder before we found out about the good stuff), you could chop wood for ten hours a day for six days straight to produce 1,000 lumens of light. That is the equivalent of a single lightbulb shining for 54 minutes.[2] Now obviously, your fire would last a lot longer than 54 minutes putting out much less intense light than a lightbulb, but the lumen is just a frame of reference for the amount of energy we're talking about.

In 1743, Reverend Edward Holyoake noted in his diary that his household had spent two days making 78 pounds of tallow candles—that they burned through in six months. If you set aside a whole week each year to spend sixty hours devoted only to making candles, you could burn a single candle for just two hours and twenty minutes each evening. That means you're going to bed at like six-thirty p.m. in the winter, and you're all huddled around a single candle before you do.

Then we graduated from stinky, smoky tallow candles made from animal fats to the gloop from dead sperm whales, which made a much better candle, but it would cost you well over $1,000 (in today's dollars) to burn a single candle each night for a year. So people weren't just keeping the lights on to party.

And then ... the lightbulb. Those sixty hours you spent chopping wood would now buy you ten full days of bright, continuous illumination instead of 54 minutes.

By 1920, that sixty-hour week of labor got you five months' worth. In 1990? Ten years. And now, you can buy fifty-two *years* of illumination for the same price ... and LEDs are only making it cheaper.

And basically everything has followed this path of getting cheaper and cheaper and cheaper. The first commercially available microwave oven in 1947 cost between $2,000 and $3,000. The first one built for consumers in 1955 cost $1,300. Ten years later, countertop microwave models were being sold for $495. And today? More than 90 percent of American households have one, and you can buy a basic model at Walmart for fifty bucks.[3]

According to the Bureau of Labor Statistics, prices for televisions are 99.19 percent lower in 2022 versus 1950, accounting for inflation. In other words, a television costing $1,000 in 1950 would cost eight dollars in 2022. In 1997, the Sharp and Sony corporations introduced their first 42-inch flat-screen televisions, which cost about $15,000 a pop. By 2019, a 43-inch LCD TV could be purchased at Walmart for $148. That's absolutely nuts! How did they even make it so cheap??

But it's not just that things are getting cheaper, we're also getting richer! Which means we have to spend fewer hours working to afford the same things. This is known as the "time price:" how many hours you have to spend to afford a particular thing. It's a great way to compare affordability over centuries and across currencies. On an average salary in 1952, you would have had to work 213 hours to afford an air conditioner that cost $350.* That was a shitload of money back then. By 1997, the price of that air conditioner had dropped to $299, but that only equated to 23 hours of work for the average worker—a 90 percent decrease.[4] And the air conditioner is like a million times better. A dishwasher went from 140 hours of work in 1954 to just 28 hours in 1997. A fridge? From 333 hours in 1958 to just 68 hours in 1997.

And it's not just appliances, either. *It's literally everything.* In 1960, it took the average Indian person seven hours of work to acquire the rice they needed for a day. Today? They can earn enough for that rice in under an hour. The time price for a comparable supply of wheat in Indiana is just 7.5 minutes.

We have to spend far less time working to afford the things we need to live, which means we can afford a whole bunch of things that make living easier. Things that were previously luxuries have become the norm. Even most of the poorest people in America (who aren't homeless) have TVs, fridges, hot water, electricity . . . things most people in poverty in

---

* That's in real 1952 dollars, not adjusted for inflation.

our country fifty years ago wouldn't have dreamed of. Not to mention smartphones and laptops and high-speed internet.

How is this even possible?? Well, technological advancement, sure. But the post-World-War-II global order means worldwide markets keep prices low due to increased specialization and competition (while leveraging dirt-cheap labor in developing economies). Mix that in with a magical anomaly of pirate-free global shipping corridors operating at the highest possible efficiency and you get everything everywhere all at once. These magical shipping corridors and massive container ships along with dirt-cheap fuel costs mean we can ship anything anywhere for literal pennies. Shipping cost is now a rounding error for products where the materials come from twelve different countries (an upsettingly large amount from China) before those pieces get assembled in Taiwan, and the next piece gets added in Singapore, and the next piece in Malaysia, before it gets sent to California, put on a truck to a massive Amazon distribution warehouse in Tennessee, before it gets put on another truck to another smaller warehouse, and then finally to your door for like, ten bucks.

And this crazy globalization stretching to every different country ready to provide that cheap labor or cheap material input means basically *everybody* has gotten richer.

In 1820, there were 757 million people living in extreme poverty. But the entire global population had only just reached a billion. Which means more than 75 percent of the world lived in what we consider extreme poverty (they just considered it daily life). Today, the number of people living in extreme poverty is lower, at about 685 million—except, we've multiplied the population by *eight*. We now have eight billion people and only 8.5 percent of them are living in extreme poverty. More than a billion people have been lifted out of poverty—since 1990.

More people have education, electricity, clean water, steady food supplies; there are fewer deaths from wars, from famine; fewer everything bad and more everything good! Lower infant mortality rates and longer lifespans. And there are soooo many different kinds of ranch dressing! We are at the absolute pinnacle of human existence!*

But wait—there's more! It's not just that we're getting richer and that things are getting cheaper and more plentiful (as wonderful as it is to have products from eighty-five different countries in the grocery store). It's that all these incredible products that so many more people can afford have taken the place of so much of the work we used to have to do . . . manually.

Before the 1950s, most housewives had to wash the family laundry by hand, and that could take an entire day. By 1961, 70 percent of US households had a personal washing machine.[5]

Before most people had electric ovens and stovetops in their homes, starting around the 1920s, you had to start a wood-burning fire every time you wanted to eat. Or bathe. And keep in mind, most of rural America wasn't fully electrified until the 1960s. Not to mention having to walk across the yard to the outhouse on cold winter mornings, because in 1940 still *half* of American households lacked indoor plumbing.[6]

Ugh, can you even imagine? You used to have to go outside to poop, and you had to like *farm* things, and wash your clothes on a washboard, and make your own candles, and make your own clothes, and knead your dough by hand . . . and rewind a little further (okay, a lot further) and you had to mill your own flour by hand. And of course, if there was a big storm or an early frost and you lost your crop—you'd just fucking die.

But now, all the food is right there in the aisle with the 39,999 other things. And we don't have to mill a damn thing or make anything at all. We can pop a Hot Pocket in the microwave and have a delicious snack that's somehow simultaneously scalding hot and ice cold in just two minutes! We have automated garage door openers, and little buttons that open the sliding doors on our minivans, and remote controls so we don't have to get up to change the channel, and blenders to blend, and freezers to freeze, and coffee makers to coffee-make, and mowers to mow, and every other time-saving machine you can imagine. And of course, big boats and trucks to cart our stuff all over the planet at breakneck speeds. And airplanes to cart *us* all over the planet at breakneck speeds. And

---

* If you want a more uplifting book than this one about how wonderful everything is, try Stephen Pinker's *Enlightenment Now* or Hans Rosling's *Factfulness*.

every one of these inventions gives us more time as humans to pursue our passions, to be creative and inventive and continue to push forward new technological advancements for our awesomely smart society! Man, we're impressive!

The Story of Everything

A human without energy is a cadaver. And technology without energy is a statue.

It's just that, somewhere along the way in the midst of all this incredible, unfathomable wonderfulness, we sort of collectively forgot that things were ever . . . hard. Every other time in human history—from cavemen to Cleopatra to Sherlock Holmes—just feels so *historic*. It feels like some backwards version of humanity before we figured out where it's really at. But whatever we're doing now is so far from any concept of "normal" human existence, it's hard to put into words.

The first modern-ish humans (*Homo sapiens*) evolved into being about 300,000 years ago. We started wide-scale farming around 12,000 years ago. Which means this last seventy-odd years is just two-hundredths of a percent of human history. If all of human history were a twenty-four-hour clock, we discovered agriculture forty-eight minutes ago . . . and this period of plenty we've come to accept as the natural state of the world? Has been going on for twenty seconds. Cleopatra just peaced out ten minutes ago.

The reason it's important to recognize how recent and abnormal all this is—the story of all that innovation and increased productivity that took us from the pyramids to the steam engine to global supply chains to the moon to everything everywhere all at once—is because the story we've been told is actually a little bit of a lie.

All our incredible human ingenuity couldn't have gotten us much past the plow or the spinning jenny without a single discovery underpinning the whole damn thing: setting fossilized plants on fire.

Because the story of everything on Earth is the story of energy.

Before we discovered fossil fuels, our energy inputs were pretty limited. We had human muscle, animal muscle, the energy from wood turned into fire, eventually waterpower around 280 BCE (such as watermills to grind flour), windmills around 644 CE, and then about 1,000 years after that, whale oil.

But fossil fuels were a million times more powerful than anything we'd ever set on fire before. It was like, "HOT DAMN you gotta be kidding me! It just keeps burning! And it burns so hot! This has GOTTA be good for something!"

What makes fossil fuels so special? Well, fossil fuels are just what they sound like. They are the fossils of mostly marine microorganisms (sorry, it's not actually dinosaurs) that lived about 540 to 65 million years ago. Their tiny little organic bodies that turned sunlight into chemical energy using photosynthesis got compressed in between the rocks over very long periods of time. And under just the right circumstances, they turned into highly concentrated, geologically stored, liquid sunlight. Pure fucking energy. Some of them turned into oil, some into coal, and some into natural gas.

Of course, almost all of the energy sources we have and use have always come from the sun. The food we eat, the animals we eat, and the first wood-burning fires of the cavemen are all just different forms of trapped sunlight. Without the sun, there would be no plants to feed us, and we would die. There also wouldn't be plants to feed our oxen and mules and other beasts that we have working on the farm and also like to eat (no animal labor or delicious steaks). Without the sun, there would be no trees to cut down to start a fire (no heat or cooked food). There also wouldn't even be wind or waterpower without the sun, but we don't need to get into that just now.

So we've spent our entire existence trying to find different forms of trapped sunlight that we can turn into energy to live. Just like every rabbit and bear and fungi and bee and plant and amoeba. And the more energy you can store up (like a squirrel hiding acorns for winter), the safer you are, the more your species can thrive. But for the first 290,000 of our 300,000 years roaming this planet ... we couldn't store shit. Back in the day, we found our energy in meat, say, when we killed a mammoth, but the meat got rancid pretty quickly, so we always had to be killing more things. We'd hunt and gather and move along once we'd used up all the resources in a place. Then one day we ran out of woolly mammoths to eat because we'd eaten them all, and we had to figure out a new way to exist. Because you know how it goes in nature: find food or die. And that's when we figured out farming. And this is HUGE. Because it meant for the first time, we could STORE our energy. And not only that ... we could have a surplus.

Holy shit! Why didn't we do this sooner!? Now we had more energy than we needed! More than we could eat! The grain could last a whole season, and we didn't have to constantly worry about where our next meal was coming from. And even better, now just 20 percent of the population could make enough food for everyone. And the other 80 percent of people could do other stuff. Like invent pottery to store our wheat surpluses, and roads to collect the wheat surpluses, and writing and math to keep track of the wheat surpluses. Or you could make babies. Lots of babies, and we could feed them all! (Most of the time.) And then shit was pretty chill for like 10,000 years while people farmed and built cities to hold all the new people and created music and math and astronomy and pyramids and Göbekli Tepe and Ġgantija and all kinds of other crazy cultural and scientific stuff no one had the energy for when we were constantly looking for food and moving to a new forest every few weeks.

But even with all the surpluses we had, food availability—and therefore energy availability—kept us in check. It placed a hard limit on population, urbanization, technological progress, and cultural expansion. We couldn't grow our societies any more than the amount of food we could grow and store.

So those super early societies and economies way back when were quite small. You could hunt; you could farm as much as your back could handle; you could trade with the folks next door . . . and that was about it. The economies we were living in were based on completely renewable resources that we could never fully exhaust. Except the mammoths, of course. And the bison. Actually, I'm pretty sure we killed off a whole bunch of species, which is why we had to start farming. But anyway, back then, if a society found themselves with scarce resources, they started finding natural means of population control. Like wars and infanticide. That's right, if there's not enough food, you just send your son to war and let your baby starve to death. During times of scarcity in some Inuit cultures, the grandpas would just hop on an iceberg and float out to sea to let the younger generation live. Bleak. But when we've got extra calories, we make extra babies. As anthropologist Marvin Harris says in his painfully academic book, *Cultural Materialism,*

> Initially pristine states probably enjoyed improved standards of living. But the temptation to use extra calories to feed extra children was irresistible, especially since child labor on irrigated lands could be made energetically profitable when children reached age six. Within a few hundred years, standards of living would begin to fall, and the peasants could be expected to take measures to slow their rate of growth by all available techniques.[7] [read: baby killing]

Until we found those sweet, sweet fossils.

Those fossil fuels gave us concentrations of energy we had never before seen. Fuck wheat surpluses, we can burn this shit. And we could move it from place to place! While the sun is very, very hot and gives us lots of free energy every day, we can't just pop it in a barrel and ship it overseas. Even solar capturing technology today is laughably inefficient. A tree stores maybe ten years or thirty years of that sunlight energy, and when you throw a log on the fire, it's gone in an hour. By the 16th century, England was already in a wood shortage because they had deforested half the country burning every tree they could get their hands on. But oil and coal are dense. Hyper dense. Like the most energy dense thing you can imagine and still carry around. It's millions of years of sunlight in a juice we can put right into our cars. Or in a barrel. Or on a truck or send swimming down a pipeline that crosses a continent. There will never be a source of energy as easy and efficient and energy dense as good ol' crude oil.

Coal isn't quite as dense or easily transportable as oil, but boy was it abundant and (relatively) easy to get to. Basically every country in the world has coal, and it's only a few hundred feet underground. We'd spent our entire existence as a species looking for ways to store energy, and here we were sitting on millions of years of stored up energy just hiding in some rocks. And all we had to do to get it was dig a really deep hole and send dozens of men down into super dangerous mines and give them all the black lung!*

The only thing was, we loved coal so much, we kept running out. So we had to keep digging deeper and deeper mines. And they were filling with water, which meant we couldn't get the coal. And I'm pretty sure lots of people were dying down there because, well, I'm pretty sure every industry was just regularly killing people a few hundred years ago. *If only we had some sort of pump that could get the water out of these mines so we could get more coal!* Luckily, one guy had the brilliant idea to burn some coal that would turn water into steam to power a machine that could help us . . . get more coal.

Enter: The Industrial Revolution.

Now, you might think the Industrial Revolution means we started burning coal to do literally everything immediately, but that wasn't quite the case. It took us a while to get the hang of all this insane energy we had access to. The Industrial Revolution began around 1760, but fossil fuel energy use didn't exceed animal and waterpower until 1870—more than 100 years later. Throughout that whole century, most coal and other

---

*   To be clear, coal had been burning in the ovens of blacksmiths since the Middle Ages, and petroleum had been burned as a gas in China going back to like 400 CE. But for some reason, the Chinese were only using it to make salt, and while the Persians used petroleum byproducts to pave roads and make kerosene lamps, the West didn't find out what they were doing until the 12ᵗʰ century. And then it still took us another 500 years to realize what we were missing and figure out how to make the stuff. So biomass (trees) and coal were pretty much the only things we had to burn.

minerals were still mined with picks and shovels, and crops were harvested and grain threshed using animal power or by hand. Even into the early 20th century, 400-pound bales of cotton were still handled on hand trucks. And barges were pulled using thousands of humans tied together with a big rope...

because that looks super fun and not like slavery at all.

Abundant coal certainly made a big dent in iron smelting and steel production, and the steam-powered engine made railroads possible (that made towns possible that made westward expansion possible), but most of those trains were still powered by burning wood until the mid-19th century. And there was plenty of industrialization that didn't involve burning anything at all, like the spinning jenny. The spinning jenny turned textiles from a "cottage industry" to a factory industry and increased a person's fabric-making speed by 800 percent, but it was still hand-powered at the dawn of all this crazy innovation.

What this revolution in industry really meant was that people could move away from the food for the first time . . . ever. Suddenly you could have a city where there was no food or water nearby because you could just ship it over there! And as we kept inventing things, we got faster and faster at farming (so fewer and fewer people needed to be farmers) so more and more people could move to the big city! At the start of the Industrial Revolution, 90 percent of the American population lived on farms or ranches. A hundred years later, that had dropped to 72 percent. Today? It's just 2 percent.

And just like how discovering agriculture gave us time to invent math, making agriculture more efficient gave us time to invent all kinds of other shit! We started inventing machines to do literally everything! Like the cotton gin and the steam hammer and the typewriter (which was ironically slower than writing by hand at the time). This fossil energy was so insanely exciting, we focused all of our collective human energy and ingenuity on unleashing it. Anything that could be done by a machine, was. And then we electrified the factories—and the machines—so we could stay up late working with all those machines even longer. Suddenly there was no limit to our population growth because there was enough food and energy to feed as many babies as you wanted. And even better, those babies on the farm had a positive ROI! You could put those toddlers to work!

And that was all *before* we started burning oil.

Sure, coal was great and everything, but oil could go anywhere. Coal was bulky and hard to transport, and you had to have a whole car on the train full of it just to keep the train going. If you were burning wood on your train you had to have like three cars of it. Five? I don't know, I just made up that number. But, ugh, what a pain. We didn't know it yet, but at the turn of the 20th century, oil was about to change everything. It was going to power all the cars and boats and trucks and plows and everything else that was about to be invented. Screw your steamboat, my shit runs on diesel. We got better, faster, stronger . . . we became the masters of the universe!

And all this invention and ingenuity and rapidly increasing productivity is the tale of the Industrial Revolution we most often hear. It's the story we're told in school: we invented all these incredible machines to make our lives easier so we could become fully self-actualized humans and experience this blissful life of ease and convenience! One day we discovered how to cook our mammoth meat and the next, we're landing on the moon! We really are impressive!

But this narrative of exceptionalism isn't just the story of the Industrial Revolution—it's the crux of the human story.

Long before fossil fuels, we mastered fire and stonemasonry and blacksmithing and built the pyramids. We invented all kinds of technology that replaced human labor (with water labor or wind labor or animal labor). Technology made us ever more efficient so we could keep growing and getting smarter and smarter. When we hooked that first plow up to our oxen six thousand years ago, farming got a lot faster and a lot more productive. And we kept improving our plow technology as the centuries passed, making incremental gains, getting a little faster and a little better with each technological improvement. We got better at making everything we were making: bricks, candles, whatever. By the middle of the 19th century, a strong man using a modern steel spade still took an estimated 96 hours to till an acre of land. To plow an acre with a yoke of oxen and a crude wooden plow took 24 hours. With a steel plow, it took just 5–8 hours. Every time we made another discovery or great leap forward, we cemented the idea that human ingenuity was the key to our newfound efficiency. We told ourselves that we're the magicians, but the watermill, the windmill, and the oxen—all the energy we were outsourcing—was the man behind the curtain.

When the unimaginable energy of fossil fuels entered the scene, it was a whole new ballgame, baby. Once we got oil (and invented a motorized plow) it was a motherfucking moonshot. By 1998, a 425-horsepower John Deere four-wheel-drive pulling a fifteen-bottom plow tilled an acre every 3.2 *minutes*.[8] We weren't making incremental gains in shaping candles a little faster or using a better kind of animal fat . . . we had gotten rid of the candles altogether. We didn't have to spend 96 hours tilling an acre of land. Or even eight hours or even five *minutes*. The energy wasn't in the technology—it was in the oil.

Most of our impressive "human innovation" over the last 100 years wasn't human innovation at all. Okay, all the machines we invented and electricity and the internet are admittedly pretty cool. We have robots performing surgery, and ChatGPT writing our history homework, and a single, highly specialized machine that can pick, separate, and bunch 4,000 radishes per hour. But those massive gains in productivity that accompanied the Industrial Revolution and the two hundred years that followed aren't a mystery. The machines don't do anything without the fucking oil.

Our lives are wonderful and easy and convenient because machines do basically all of the jobs we don't want to do. We don't have to do laundry (though people still complain about "doing" what little laundry we have to do) or spend a week making candles or chop wood or wash dishes or walk places. We don't have to make clothes, and we certainly don't have to pull barges. All of us have lived our entire lives in the only blink of an eye of automation that has ever existed. And every day we create more machines to replace as much human labor as possible. The fewer humans the better! Since this all sounds pretty great, you're probably wondering what my point is. The point is that we've become blind to the amount of energy we use just to live every day. We've forgotten about the machines that do literally everything and what keeps them running. We've forgotten that the fuel we're burning won't last forever—it can't. And the more advanced society gets, the faster we burn it.

A hunter-gatherer society used about 20 gigajoules of energy per person per year, which works out to about 13,000 kilocalories per day.[†] Most of this was just burning wood for fires and eating food so you had energy to go hunt more food. In an agricultural society, that increased to 60 gigajoules per year (about 40,000 calories per day), which was also burning wood for fire, but now we're also feeding our oxen that we need to plow our fields. Today, in a fully industrialized society, we're looking at 150–300 gigajoules per year.[9] That's 100,000–200,000 calories per person per day.

Daily per capita energy use by social structure
Exosomatic/endosomatic energy consumption

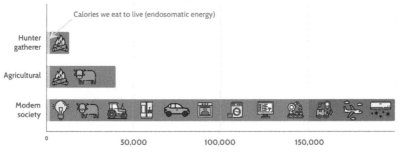

Data Source: "Human energy use, (exosomatic/endosomatic), Environmental Justice Organization

---

† In America, we use "calorie" and "kilocalorie" interchangeably. When that slice of pizza has 285 calories, they're really talking about kilocalories. So hunter-gatherers used the energy of 45 slices of pizza per day.

But not only are we using 15× more energy per person to power our society, we've increased the number of persons by eight-fold in the last 200 years. Here's a little chart to put these numbers into perspective.

On a kilocalories basis, the United States burns through 24 quadrillion every year. Dang, that's a lot of pizza. Oh, did you not read that last footnote? Read the footnotes. They're funny and informative! A slice of pizza has about 285 calories, so if our country was powered by pizza that would be about 10.5 trillion pizzas a year or 31,236 large pizzas per person. Including babies. And your grandpa.

All-in-all that's 72 million calories of energy use per person per year. Of course, we don't eat the vast majority of those (sadly). At 2,000 calories a day, we can only shove about 730,000 calories of pizza in our faces annually. Which means—and this is the important number—we're spending **100×** the energy we need to feed ourselves just driving around and ordering crap on Amazon and turning on lightbulbs and turning on the heat in September while wearing a t-shirt and shorts . . . every single day. This is called exosomatic energy. It's energy we don't put in our bodies, but that we use to fuel the lives we lead. You can see the ratio of exosomatic to endosomatic energy usage in the chart above. No matter which type of society we have, we need the same amount of calories to live (more or less). The only thing that changes is how many calories we burn doing all the other things we do to support our existence.

Another way to put all this exosomatic energy into perspective is to think about how much work it would take a human to do it. For every machine we made that made our lives easier, we were replacing the labor of a human: making a scarf, picking a radish, whatever. All that human ingenuity driving massive progress (and massive profits) was mostly just *b*illions of years of work that humans didn't have to do fueled by a finite resource that we're sucking out of the ground ten million times faster than it was put there.

Now does it sound crazy to you?

How many licks to the Tootsie Roll center of a Tootsie Pop?

**The mansion of modern freedoms stands on
an ever-expanding base of fossil fuel use.**

—Dipesh Chakrabarty
The Climate of History

How many humans was it really? How many years of human labor do you think we replaced when we started burning sunlight trapped in the ground ten million times faster than the sun put it there? Excellent question.

The human value of a barrel of oil is something that's hotly contested in oil forums on the internet[10] (yes, these are a thing). But some back-of-the-napkin math will help put the power of oil into perspective for you.

For a little opening context, one barrel of oil contains 1,700 kilowatt-hours of energy. The average American house in 2021 used 10,632 kilowatt-hours annually, or about 6.25 barrels of oil.[11] There are a lot of different ways to quantify energy, and I'll use a lot of them throughout the book, but here's some math to get you familiar with a few:

» A typical, 42-gallon barrel of oil contains 5.8 million BTUs (British Thermal Units), or 6.1 billion joules of energy

» 4,184 joules = 1 kilocalorie, so 6,100,000,000 joules/4,184 = 1,457,934 kilocalories

» So, one barrel of oil has 1.46 million kilocalories[‡]

Now for the squishy part. Over an eight-hour work shift, an average, healthy, well-fed individual uses about 120–750 kilocalories per hour to do their job. This number is lower if you're working at a computer, and higher if you're digging a ditch.

At 2,000 hours per year (assuming you work a forty-hour week and take two weeks of vacation) at the highest end of the spectrum, this is 1,500,000 kilocalories per year—about the same as a barrel of oil. At the other end of the spectrum sitting at our computers, we use just 240,000 kilocalories per year, which would mean there were six *years* of human labor at that intensity in every barrel.

Another way to look at this is from human output (how much energy can a human *create* instead of how much we use. These estimates range from 750 BTUs in peak physical condition (7,733 human hours per barrel), but closer to 350 BTUs for sitting at your desk (which would be 16,571 human hours per barrel). But it doesn't really matter. No matter which way you slice it—calories or BTUs, sitting, standing, digging a ditch, whatever—each barrel of oil does somewhere between one year and nine years of human labor.

Is that a big difference? Yes. It's almost an order of magnitude. But it *still* doesn't matter. Here's why.

The US population in 2021 was about 337 million people. In 2021, we used 97 quadrillion BTUs of energy in total.[12] That's our 10.5 trillion pizzas, and it includes renewables and nuclear and everything. Of those 97 quadrillion BTUs, or "quads," 76.9 came from fossil fuels. In every quad, there are 180 million barrels of oil, give or take. So that gives us a total of 13.85 billion "barrel of oil equivalents" or BOEs. This is a handy way to talk about fossil fuel energy coming from multiple sources.

_____

‡ Or 5,115 slices of pizza.

SO . . . in 2021, 337 million people burned through 13.85 billion BOEs, which means each human in the United States—including babies—used about 41 barrels of oil. This isn't just the car you drive or your gas stove, it's all the energy it takes to run the whole country. It's the truckers that drove your radishes from California to Virginia. It's the machines that helped fertilize and plant and harvest those radishes. It's the fertilizer itself, which is also made from fossil fuels. It's the forty-seven Amazon packages that got dropped on your doorstep last week. It's the lights in your home, the lights everywhere—it's everything.

Of course, some people use less. If you're a frequent flyer, your number is higher. If you live in a city and only take public transport, it's lower. If you're a vegan without a car who doesn't do air travel, it's even lower. But 40 BOEs a year feels like a good jumping off point. So, if we assume each barrel of oil does the work of a single human for somewhere between one and nine years, then this implies a "fossil fuel slave subsidy" of around 40–360 human years per person.

TL;DR: I know that was a lot of math, but just to clarify: Each year, every person in the United States has the work of somewhere between 40 and 360 human slaves done for them by machines, powered by oil, in order to support their very existence. Every. single. person. And there are 337 million of us. That's a lot of slaves.

Now for the even more insane part. How much oil costs. Like nothing. At the time of writing, a barrel of oil cost $85.

Even at the laughably low minimum wage of $7.25 and two thousand hours a year, a worker would cost $14,500 for a year. If there are five human years in every barrel? Well, 5 × $14,500 = $72,500 per barrel. Start paying a living wage, and you can triple that.

And that doesn't even include the cost of the environmental harm it's doing.*

Are your eyeballs bleeding yet? Please take a moment to rest from all that math. There's just a little more coming.

---

* So much more on that later.

So if we know how valuable oil is when it replaces a human picking a radish with a machine that does it in a tenth of the time, then why the hell is it so cheap?

Oil is so cheap because it's priced based on the cost of extraction: how much it costs to get it out of the ground. And how much demand there is, and messed up shit like the war in Ukraine. But that's incredibly misleading. The real cost of oil is the human labor it's doing in our place. Because if we didn't have oil, that's what it would cost us to get all that work done. Every single product you buy in our magical world of global shipping and cornucopia of DiGiornos is drastically underpriced when you think about how many human workers it would take being paid an average human wage to make that same product without the machines that power our existence—and how much oil they burn in the boats and trucks that deliver it to your door.

Now let's zoom out a bit further. Just a little more math, I promise.

The US Energy Administration says that from 1950 to 2021 the entire country used 4,775 quads of fossil fuel energy. That's 4,774,887,000,000,000 BTUs. I just wanted to write out all the zeros so you could see what a quadrillion looks like. There are 5.8 million BTUs in each barrel (5,800,000), so in the last seventy-one years, we've burned through 823.4 billion BOEs. Just in America.

If each barrel of oil has five years of human labor give or take, and the average human works about forty-five years of their life conservatively, it takes nine barrels of oil to match the energy of an entire human's working life. So, divide that 823 billion barrels by nine and you get 91.4 *billion* fossil fuel laborers devoting their entire theoretical lives to keeping your fridge cold. And that's just since 1950. The first oil well was drilled almost a hundred years *before* that.

I know this math is very squishy, and you can argue it a million different ways, but it's back-of-the-napkin to prove a point. I don't really care if it's 9 billion or 90 billion or 900 billion. The point is, we've built our glittering, fast-paced economy with every convenience you can imagine on a seemingly boundless supply of fossil fuel workers doing all the jobs we don't want to do a hundred times faster than we can do them—and we pay almost nothing for the insanely valuable, irreplaceable substance that powers the whole machine. *Now* does it sound crazy to you?

America isn't the richest country in the world because we're the best at innovating things. We're the richest country in the world because we've used far more fossil fuel slave labor than any other country on Earth. And it's true for every single country that industrializes. The more jobs you give to machines instead of people, the more fossil fuels and electricity you use, the more economic growth you see. Per capita electricity consumption correlates almost perfectly with economic growth. The more kilowatt-hours of energy you use, the more your economy grows, the richer everyone gets. And developing countries moving from their subsistence farming lives to developed, industrial economies like ours are just going to continue to use more and more energy as they get bigger and bigger. Because that's what we do. Global energy demand is already predicted to grow by 30 percent before 2035—and there are still *1.2 billion* people who don't even have electricity.

Sooooo . . . what's not to love? Oil makes life easier. I drive my car everywhere and look at my phone while the TV plays in the background and my laptop sits open on my desk, and my fridge keeps my food cold, and I fly home for Christmas every year . . . who cares? Everybody's richer and nobody has to poop outside. You're not making a great point here, Taylor. Except that one tiiiiiiny, eensy weency problem that no one's talking about. And I don't mean climate change.

The Boy Who Cried Peak Oil

We're never going to run out of oil. What the world is going to run out of, indeed, what the world has already run out of, is the oil you can afford to burn.

—Jeff Rubin

I remember when I was in middle school back in the '90s learning about "peak oil" and how close we were to it. But then one day, everyone just stopped talking about it. Why? Because we kept finding more oil, I guess.

But what is "peak oil" exactly?

Peak oil is the time when the maximum rate of global oil production is reached. After this point, we'll be getting less and less oil every year, until eventually, there just isn't any left. Everybody is aware that oil is finite. It was created tens of millions of years ago and it will take millions more years to make any more. The only part that's up for debate is how much is actually left.

Fifteen or twenty years ago, people were talking about this a lot. Actually, people have been talking about this for more than a century. Back in 1919, David White, chief geologist of the United States Geological Survey, wrote of US petroleum: "The peak of production will soon be passed, possibly within three years." Lolz. In 1956, this dude M. King Hubbert predicted that we would reach peak oil in the year 2000 at 12.5 billion barrels per year, or 34 million barrels per day. To put how wrong he was into perspective, in November 2018, world oil production reached 102 million barrels per day—about three times the supposed "peak."

The problem (or the solution, depending on how you look at it) is that every time someone predicted peak oil, we found more oil, or we found ways to extract oil we couldn't get to before. So the amount we were pulling out of the ground every year just kept going up, and it was like "the boy who cried peak oil" time and again.

But that doesn't mean we're out of the woods.

First, we're really not finding new oil reserves anymore. At least not like we used to. Historically, new discoveries accounted for most of the additions to global reserves. Without new additions, the oil we have declines at a rate of about 7 percent every year. That's pretty fucking fast. Forty percent of *all* the oil we've ever found has come from just 900 oil and gas fields. There just aren't as many places to look. Don't get me wrong, we're still discovering oil, but at a much lower rate, and in places where it's a lot harder (and more expensive) to get out. Seventy-seven percent of new discoveries in 2022 were deep water reserves.[13]

Second, one of the main ways we avoided hitting peak oil ten years ago wasn't finding massive new oil reserves, it was developing new technology to get oil that was too hard to get before. The US—because we're so flippin' rich, and we love oil so much—has been at the forefront of these technologies.

There's an argument to say we hit peak oil nearly fifty years ago as a country . . . because we did. Conventional oil production peaked in the US in the 1970s and has been steadily declining ever since. We've made up basically all of that decline with technological improvements—and one in particular.

US Oil Production vs. Hubbert Upper-bound curve

Comparing lower 48 states oil production from 1900–2020 with M. King Hubbert's upper-bound production estimate from 1956.

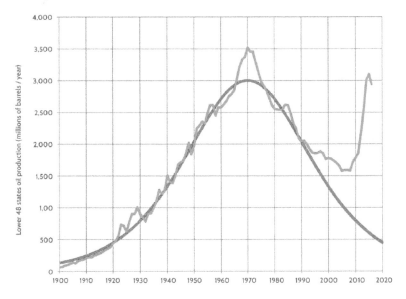

See that green line (or if you're reading in black and white, the line that jumps up around 2008)? See how it definitely peaks in 1970 and then drops off for decades? Welp, I'm sure you've heard of "fracking" and how it's terrible for the environment and how it poisons drinking water and causes earthquakes, and you know, minor things like that. But environmental arguments aside, it lets us get at a LOT more oil we couldn't get before.[†] Hydraulic fracturing (aka fracking) isn't completely new (the first successful commercial application was in 1950), but we've gotten better and better at it over the years. Hey look! We do innovate things!

What is fracking exactly? Well, traditional oil deposits (think Beverly Hillbillies oil) exist in liquid form in highly porous and permeable rock underground, like sandstone. Because the rock is porous (full of holes) and permeable (easy to flow through) all you have to do is drill into it, and BAM! We're rich! Texas Gold comes flowing out of those holes with very little work, for decades.

---

†   Once again, refer to that green line jumping way up around 2010.

It sounds weird, because like, how does a rock hold liquid? But sandstone is the sponge of the rock world. It holds lots of liquid. Up to 35 percent of sandstone is just holes. Holes often filled with that sweet, sweet crude oil.

But there is also a whole bunch of oil and gas that's trapped in less permeable rocks, like shale. In the past, these little pockets just didn't have enough juice to be economically viable as an extraction source. So, while we knew they were there, they didn't count towards our reserves.

Enter: fracking. Fracking means we take a bunch of water mixed with sand and other chemicals and blast it into the rock until it cracks, which releases all the tiny pockets of good stuff trapped inside. If you've heard of the term "horizontal drilling," it's also related. Here, instead of just drilling down, you drill down, then across—often up to two miles across—and then frack the shit out of all the rock. Since layers of rock underground tend to be largely horizontal, this means more bang for your fracking buck. You can also drill horizontally for traditional oil, but its biggest value-add is in the fracking space, most notably in natural gas shale reservoirs. Why? Because the gas is trapped in tiiiiiny pockets in a rock with very low porosity and permeability—fewer, smaller holes and harder to get through.

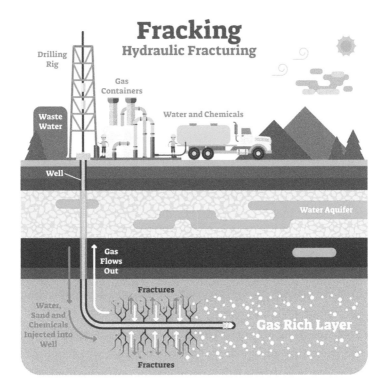

So once we could get at all that natural gas and crude oil trapped in the shale, it opened up a whole new world of oil production. Improved fracking technology is why the US became a net exporter of crude oil in 2019, and why oil in our country (known as West Texas Intermediate Crude, or WTI) is consistently cheaper than global oil (known as Brent Crude).

Okay, great! So other than causing earthquakes and releasing tons of methane and poisoning drinking water, fracking sounds awesome! We can get oil we couldn't get before! But there's one more catch.

US oil production peaked in November 2019 at 13.0 million barrels per day (mmb/d), then dipped during the pandemic, but is climbing back toward its original peak at 12.7 mmb/d as of March 2023.[14] So what's the catch? We still haven't hit peak oil!

Well, those numbers aren't telling the whole story. Sure, we're producing more barrels than ever before. But the oil we're getting now isn't anything like the oil we were getting back in the '70s. In fact, 60 percent of our "oil" production is actually non-petroleum, mostly in the form of natural gas liquids (NGLs). So only 40 percent of the oil we're getting out . . . is actually oil.

That's the thing about shale plays. Shale and other tight oil deposits have a much higher percentage of natural gas than crude oil. While we also use and love natural gas, it doesn't serve the same purposes as crude. It can't be put in a barrel, or put in an oil tanker, or sent down a pipeline without being highly pressurized. Oh, and it's super explosive. It has a higher energy density than crude oil, but we can't turn it into gasoline for our cars or jet fuel for our planes or diesel for all the trucks and boats that power the global economy. And the NGLs we're getting are an even more niche type of fuel.[‡] NGLs now account for 30 percent of US oil production, but we can't put those in our cars either. They're pretty much only good for making plastic bags, anti-freeze, and detergent, along with propane for our grills and butane for our lighters.

Tight oil, while it *is* petroleum, also still kind of sucks. It's only about 93 percent as energy dense as the good stuff, which doesn't sound like a big difference, but it is. Tight oil also doesn't have the compounds we take from normal oil to make diesel. So in a world where the US lives on tight oil, there's no diesel production, and we have to import the heavier crude oil from somewhere else (while it lasts). And just to be clear: not a single city in the world could survive without the large-scale diesel transportation that brings in every single bite of food they eat. Diesel powers all the trucks and all the machinery and all the farm equipment and pretty much everything. Diesel is truly the workhorse of the modern global economy.

And the *even* worse part is that in 2021, tight oil accounted for 64 percent of US crude oil production, up from 52 percent in 2015 and basically nothing in 2010.

Okay, but whatever. We're still finding lots and lots of it! Let's frack until the wheels fall off!

Well, there's another thing.

---

‡ Not to be confused with LNG, which is liquefied natural gas. This is what happens when we pressurize the regular stuff so we can put it in tankers and pipelines.

The oil and gas deposits that we can access with fracking just aren't as big as the old school ones. These deposits are called "tight oil" for obvious reasons. While a traditional oil well will be productive for twenty to thirty years on average (steadily declining over time), a fracked well will give up almost all of its reserves in the first *two years*.[15] It's *exponential* decline. Bummer. Which means if you want to keep up production, you have to keep making new wells. All. the. time.

According to petroleum geologist and expert on US shale plays Art Berman, if we stopped drilling new wells in the US today, our total oil production would drop by 40 percent in the first year. So what's the problem? Drill, baby drill! Well, we will. But Berman likens fracking technology to having a bigger straw. We can suck more out of the ground a whole lot faster—and we've gotten really good at making extra-big straws—but it doesn't increase the amount that's available. The only thing that does that is time. Lots and lots of time. And after fracking sucks the ground dry . . . we're out of ideas.

As our straws start slurping at the bottom of an almost empty glass, there's an important concept to understand: EROI, or Energy Return on Investment.[§] For any energy we get, whether oil or gas or solar or whatever, we have to spend energy to get it: energy building the oil well or the solar panels; energy turning the crude oil into gasoline; energy figuring out where to go look for the energy. The EROI of our old Beverly Hillbillies oil was huge. You'd stick a fork in the ground and kick your feet up for thirty years. When we first struck oil, the EROI for finding it was 1200:1.[16] That is to say, for every calorie we spent finding an oil deposit, we'd get 1,200 calories back. That's a pretty incredible ROI. By 2007, that number had dropped to just 5:1. In terms of production (energy spent getting the oil and turning it into usable energy), it fell from a peak of about 24:1 in 1954 to 11:1 in 2007. Now, that's in the US. The current EROI of global oil production is 17:1.[17] But an even more messed up thing happens when the EROI falls below 10. The price of oil starts increasing in a non-linear fashion. It's inverse and exponential. So as that number gets closer to 1 (we have to spend a unit

---

[§]   EROI is messy and calculated differently depending on who you talk to, but the numbers are scary no matter which way you slice it.

of energy to get back maybe 1.5 or 2 units of energy) the price of oil will get *exponentially* higher.

And another crazy fun fact is that the US oil and gas industry already *uses* the most energy of any industry. We're spending 5–10 percent[18] of all those 72 million calories per person per year just looking for energy that we can turn into energy.

So if tight oil deposits run out more quickly . . . and we're now getting the majority of our oil from tight oil . . . and most of the tight oil deposits are lower in quality . . . and we're using way more energy to find the energy we do get . . . what does that say about future oil production? About future oil prices?

It's not looking good.

General consensus is we have about fifty years left based on the reserves we know we have and current demand. Fifty years. I'm not sure how to stress to you that fifty years is not that fucking far away. The world has used half of the oil it's EVER used since 1995. Just in the past twenty-eight years. And no one is predicting a global slowdown.¶

If anything, we're going to keep using more and more as countries in the Global South develop further and look to have the same luxuries we've all been enjoying for the past half century like cars and TVs and refrigerators and DiGiornos.

Now imagine the entire world didn't just get a fridge but started using 100× exosomatic energy like the good ol' U-S-of-A . . . all eight billion of us. And it'll be more like ten billion by 2050. Does it still feel like oil is going to last fifty more years? Or is it maaaaaybe starting to feel like we might run out a little sooner than predicted? Oh, and destroy the entire planet in the process.

No matter which way you slice it, it's gonna run out. "Peak Oil" isn't a theory. It's an empirical fact. At least what's usable to us. There will always be some oil trapped in insanely difficult and cost-prohibitive places. But then technology will improve, reserves that seemed impossible will become possible, and we'll certainly keep finding tight oil and

---

¶   Here is a fun countdown where you can watch us burn through our reserves in realtime! https://www.peterleeds.com/oil-clock.htm

squeezing every last bit out of the ground for decades more. That fifty years could turn into seventy, sure.

The point of all this isn't to predict a date for peak oil or a date we're gonna run out of oil. The point is that there is a tipping point where whatever's left in the Earth will become too expensive to extract to be economically viable. And the even more important point is that as we approach *that* point, the price we pay will increase exponentially. We *already* can't access nearly as much, it's *already* harder and more expensive and energy-intensive to get, and we're *already* getting an inferior product that runs out exponentially faster when we do. We have to keep spending more and more money and energy just to get the oil we have left out of the ground. And as it costs more to extract and becomes more scarce, everything that uses it will cost more. Which is literally everything.

What we're dealing with isn't the end of oil ... it's the end of abundant, cheap oil that our entire globalized society depends on.

And we haven't even gotten to the climate yet.

Hold onto Your Butts

Hold onto your butts.

—Samuel L. Jackson
Jurassic Park

But we're not gonna run out of oil because we keep getting more efficient! We do it every year!

Great plan! And very true! But the messed up thing about getting more efficient is that it might actually make things worse. Because the more efficient we are, the more we grow, the more energy we use, so efficiency never actually makes us more efficient. It's called the Jevons Paradox.

William Stanley Jevons was a British economist who realized back in 1865—almost two hundred years ago—that increasing the efficiency of energy and material use often leads to more, and not less, consumption of the energy or raw material. Why?

Well, technological progress increases efficiency. That's great! Now we use less energy to do the thing that used to take more energy. The problem is two-fold: because it's more efficient, it now becomes cheaper, so more people use it (and therefore use more energy). The other piece is that when an efficiency is created, that means a human has more time (or money) on their hands, which they then spend using even more energy, like a flight to Hawaii or watching twelve hours of Netflix (one of those uses far more energy than the other).

We've spun the narrative that becoming more energy efficient as a society is the answer to our fossil fuel problems. But if all those increases in productivity made us use less energy, we wouldn't continue to have that 1-to-1 coupling of energy use and economic growth. People would keep getting richer, but our energy use would stay flat because we've gotten so much more efficient! But that has yet to happen. And once again, brilliant economists have just decided to ignore this very obvious, provable fact that some dude figured out two centuries ago. Or as told in another highly accessible passage from Marvin Harris:

> Cultures have not generally applied the increments in techno-environmental efficiency brought about by the invention and application of "labor-saving devices" to saving labor but to increasing the energy throughput, which in turn has not been used to improve living standards but to produce additional children.

That guy's a real joy to read

Global GDP vs. energy consumption, 1965–2015

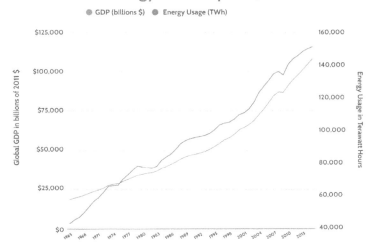

Data Source: BP Statistical Review of Energy, 2022; Our World in Data

The good news is, in the past decade, a handful of fully industrialized countries have started to decouple growth and energy usage a bit. Some more than others.

From 1995 to 2019, Sweden managed to grow their per capita GDP by 54 percent while reducing consumption-based energy per capita by 12 percent.[**][19]

The United States saw per capita GDP rise 45.5 percent over the same period, with consumption-based energy increasing just 2 percent. The UK, Germany, Denmark, and Switzerland have also all managed to grow while keeping energy use flat or even declining in some cases.

While this is somewhat encouraging, I hate to burst your bubble . . . it still doesn't change the amount of *absolute* energy we're using. Even if we're not growing energy usage *as* fast as GDP, so we've achieved "decoupling," it's still growing. Using less energy per dollar of GDP growth is great, but if you're still adding more people and a whole bunch more dollars of GDP, it doesn't really solve anything. And developing countries have a looooong way to go before they can reach even a hint of decoupling. Which means we still need a lot more energy to keep up.

But we're gonna get all the new energy from renewables! The future is here!

Hooray! Unfortunately, any big chunk of energy that we've replaced with renewables or even natural gas has, once again, been more than offset by growths in absolute energy use. And while we've traded out a lot of dirty, dirty coal for clean(er) natural gas, (mostly because of the shale natural gas boom from fracking), renewables still make up a tiny percentage of our overall consumption.

[**] "Consumption-based" means they are accounting for goods produced in another country but imported and consumed in the country at hand. This helps adjust for rich countries that no longer do most of their own manufacturing, so their energy costs can appear to be going down, while their real energy usage is just being shifted to developing countries.

Share of energy consumption by source, United States

To convert from primary direct energy consumption, an inefficiency factor has been applied for fossil fuels (i.e. the 'substitution method').

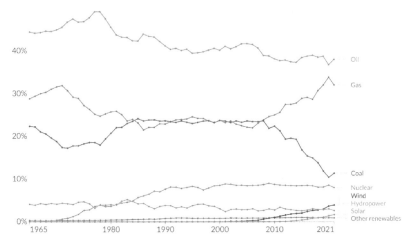

Data Source: BP Statistical Review of World Energy (2022) | OurWorldinData.org/energy | CC BY

The renewables are helping to curb increases in fossil fuel use, but they're not decreasing the amount of fossil fuels we use. At all. They're just helping us continue to grow. Because man, are we addicted to growing. I'm sure renewables will overtake fossil fuels one day. I mean, they're gonna have to. But we are nowhere near that day.

But the electric car revolution is here! As soon as we all stop driving gas guzzlers, it'll be fine.

Please continue to hold onto your butts: electric cars are not going to save the world. Are they better? Sure. I think so. Maybe? Well, it depends on what you're optimizing for. But there's a solid argument that electric cars are worse for the planet than your conventional gas guzzler. And they still use a boatload of fossil fuels to produce. In fact, electric cars create an estimated 80 percent more emissions during production than fuel-powered cars.[20]

How is that possible, you ask? Well, electric cars, much like regular cars—much like everything—have to be built. But the problem with electric cars is that they take a LOT more energy and a lot more materials to build than regular cars.

EVs are made mostly with aluminum instead of steel, for weight purposes, and aluminum takes *six times* the energy to make compared to steel. Aluminum doesn't just grow on trees; you have to mine bauxite and then refine the bauxite—which we need oil to do. EVs take less aluminum than steel by weight, but you can assume at least double the carbon intensity just to make the frame.

But the real energy suck, ironically, is the battery.

To make a battery for a single car you need to refine 25 pounds of lithium, 60 pounds of nickel, 44 pounds of manganese, 30 pounds of cobalt, 200 pounds of copper, and 400 pounds of aluminum, steel, and plastic. And all these materials get mined in one place and refined in another (at super high temperatures) and then shipped to another (on oil-powered boats). Most of the lithium today is mined in South America, but China has 60 percent of the processing capacity.[21]

Just to give you a picture, here is a fun map of all the places we get the materials to make a car battery:

Supply chain of raw materials used in the manufacturing of light-duty vehicle lithium-Ion batteries

"NMC" is shorthand for "nickel manganese cobalt oxides," the primary battery material in lithium-ion batteries.

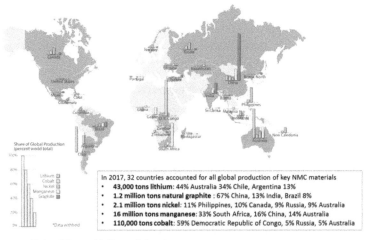

In 2017, 32 countries accounted for all global production of key NMC materials
- **43,000 tons lithium:** 44% Australia 34% Chile, Argentina 13%
- **1.2 million tons natural graphite :** 67% China, 13% India, Brazil 8%
- **2.1 million tons nickel:** 11% Philippines, 10% Canada, 9% Russia, 9% Australia
- **16 million tons manganese:** 33% South Africa, 16% China, 14% Australia
- **110,000 tons cobalt:** 59% Democratic Republic of Congo, 5% Russia, 5% Australia

Data source: National Renewable Energy Laboratory

Notice how the United States makes literally none of it? These vital raw materials get shipped all over the world at the mercy of our incredibly fragile global shipping corridors that are also dependent on—you guessed it—super cheap, abundant oil!

And of course, right now, if you plug in your car, it's connecting to an electric grid that's still powered mostly by fossil fuels. Let's check out some number-crunching from Peter Zeihan's recent bestseller about the impending collapse of globalization, *The End of the World Is Just the Beginning*:

> A typical 100-kilowatt-hour lithium-ion battery is built in China on a largely coal-powered grid. Such an energy- and carbon-intensive manufacturing process releases 13,500 kilograms of carbon-dioxide emissions, roughly equivalent to the carbon pollution released by a conventional gasoline car driving 33,000 miles. That 33,000-miles figure assumes the Tesla is only recharged by 100 percent Greentech-generated electricity. More realistically? The American grid is powered by 40 percent natural gas and 19 percent coal. This more traditional electricity-generation profile extends the "carbon break-even" point of the Tesla out to 55,000 miles.[22]

So, sure, you'll probably (hopefully) drive your EV for more than 55,000 miles, and you can say "Hey! This is *equally* as bad for the environment." And then the longer you drive it, the less bad it gets. And you're not wrong. The Union of Concerned Scientists (lol) estimated that over the lifespan of an EV, it generates about half as much emissions as a conventional car. So is it better from an emissions standpoint? Sure. But are we using more fossil fuels in making the car? Yup. Does it even out over time? I think so. Are there a whole bunch of other reasons why electric cars still can't save the world? You bet.

Of course, those numbers get better and better as we switch to more renewable sources powering the electric grid that powers the car. And if one day, we switch the *entire* electric grid to renewables, the whole fossil-fuel-charging part of the argument will eventually be moot!

Sure! Eventually! But right now, renewables are only providing 28 percent of electricity worldwide.[††] Not only do we need to switch the entire grid to renewables to power all the things we already have plugged in, but if everybody switched to an electric car today, we'd need to *double* the electricity generation capacity of the planet to handle the extra load. So, you know, that's not really feasible anywhere in the near future. And in the meantime, where do you think we'll be getting all that electricity? I'll give you a hint: it rhymes with schmossil schmules.

But I heard we just achieved a net-positive nuclear fusion reaction! That's a thing, right?!

I know! So exciting! This could be a thing! Sort of. So, yes, in case you didn't hear, scientists achieved a nuclear fusion reaction that created more energy than was put into it. The reaction itself created 3.15 megajoules from lasers that pumped in 2.05 megajoules. So there is a positive EROI! Except . . . 1.1 megajoules is only a third of a kilowatt-hour. Which is about half as much as it takes to dry your hair with a blow dryer. But okay, net positive, that's still good! Except . . . it wasn't net positive at all. Because they had to use 300 megajoules from the electric grid to power the reaction—or 300× the energy the reaction actually produced.

Still, very cool, could be promising! But the "decades away" many scientists are claiming for commercial fusion feels a lot more like a pipe dream to me.

And considering how absolutely massive any commercial nuclear fusion reactor has to be, and the materials, and the lasers, I don't know, maybe we should just keep to the solar and wind?

Ooh! What about deep geothermal???

Yesss, now we're talking! If you haven't heard, deep geothermal is probably the most promising new technology out there. Wayyyy more

---

[††] Remember, this is referring to electricity only, not total energy usage.

promising than nuclear fusion. Basically, we drill massively deep holes into the Earth to get at the giant nuclear core of our planet. The nuclear reactor that we live on top of is so flippin' hot that tapping into just 0.1 percent of it could supply the entire world's energy needs for 20 million years! The reason we're not already doing this is we don't have the technology to drill holes that are deep enough to get at that heat. Back in 1989, the Russians got all the way down to about 7.5 miles deep (12.2 km), but it was even hotter than they thought, the rock was strangely porous, and none of the drills they made could stand up to the literally hellish conditions down there. So everyone pretty much gave up on that idea. They literally just put a lid on the hole and walked away.

Except now, we may have invented a whole new type of drill that vaporizes rock using nuclear fusion lasers!‡‡ It's called direct-energy drilling. We use a wacky machine called a gyrotron that, since it vaporizes rocks using energy, won't have any of the issues with mechanical drill bits breaking at depth. AND they can re-purpose existing coal-fired plants to run on the new clean steam from deep within the Earth, leveraging existing infrastructure! AND deep geothermal exists everywhere on the planet! This sounds better by the second! And this isn't totally science fiction, either. A company called Quaise is hoping to re-power the first fossil-fired plant by 2028! Of course, they haven't actually drilled a single hole outside of the lab yet. And we'll need thousands of these holes to make a difference in global energy production. And each of those holes will need millions of liters of water per day to transfer the heat energy from ten miles underground. And you can guess how much energy is needed to power these millimeter-wave laser beam drills and pump all that water. And you can guess where we're getting that energy in the meantime.

Okay, so we'll keep working on that, and we'll expand renewable energy capacity as fast as possible. We gotta build wind farms and solar and hydro and everything! We can do it! Planet Earth Unite! We'll spend trillions of dollars renewableizing the whole damn world and we'll never burn another fossil again!

---

‡‡   Holy shit, we really are impressive.

Sure sure sure. There's just one other thing. As much as we're blind about the amount of energy we use, we're also pretty blind as to how the sausage gets made.

I want you to pull your phone out of your pocket and think about it for a second. Think about how it came into being. You're not really sure, are you? Surely someone had to mine some copper or gold or lithium for the battery. Which countries have those? And I guess some oil went into the plastic that made the cover and the buttons? Where was that refined? How does oil turn into plastic anyway? And I guess someone else had to make the glass for the screen? And the semiconductors? And what about the technology itself? That's a whole other conversation.

All-in-all there are about *sixty* raw materials that go into your phone. That tiny thing you're always looking at. Now imagine how many raw materials go into . . . I don't know, literally anything bigger than a phone. Like, say, maybe, an electric car. Or a wind farm.

Here's a not-so-quick overview of what a handful of those materials do and which countries produce the most of them, just to give you an idea:

Plastic (a byproduct of crude oil refining) – for chip coatings, protective casings, and electrical components.
USA, Russia, Saudi Arabia, Canada

Glass – Phones use a specialty glass composed of silicon dioxide and aluminum oxide with an ultra-thin coating of indium tin oxide. But most of the glass is just silicon.

Silicon – Mainly for the processor, but represents almost 25 percent of the raw materials needed. NBD though as it's everywhere— it's nearly 30 percent of the earth's crust.

Lithium – Absolutely essential for the battery. And there is not nearly enough of it that we can access right now.
Australia, Chile, China

Iron – For all the screws as well as body parts and the batteries.
Australia, Brazil, China, India

Aluminum – Used in plates to shield the electronics from the electromagnetic radiation coming from the antenna. Aluminum is refined from bauxite.
China, India, Russia, Canada

Copper – Used more than any other metal in the phone, mostly for the cables as well as chips and printed circuits.
Chile, Peru, China, Democratic Republic of the Congo (DRC)

Cobalt – By far the most expensive raw material, cobalt has the highest energy density among metals and is absolutely critical for the battery. It's also one of the hardest to find in the world.
DRC, Zambia, China

Tin and Lead – Used to solder different components to the motherboard.
China, Indonesia, Peru

Zinc – Alloys with aluminum and copper to make microphones and speakers and in battery manufacturing.
China, Peru, Australia, USA, India

Nickel – Used in batteries, capacitors, and electrical connectors.
Indonesia, Philippines, Russia

Silver – Used in the conductive tracks of the printed circuit board.
Peru, Mexico, China, Australia.

Gold – Used for conductors, switches, buttons, RAM, motherboards, and cables.
China, South Africa, Australia, USA

Gallium – Used in LEDs as back lighting of the display or camera light. Gallium is a byproduct of mining aluminum, or sometimes zinc.
France, Kazakhstan, Russia

Indium – Used for LCD displays. This is a very rare metal.
China, Canada, Peru.

Palladium – Used for the contact surfaces between individual components.
Canada, South Africa, Russia.

Neodymium – Used for magnets in the speaker, receiver, camera, and vibration mode. It's the strongest magnet ever discovered. It's extremely expensive and extremely rare.
China has basically all of it.

Tantalum – Used as a capacitor.
DRC, Australia, Brazil.

And then ... there's the mining. Oh my god, the mining. To get all these rare earth metals needed to make your iPhone (or your Tesla battery or your wind turbines) you've got to mine them in rare earth mines. And they exist in tiny quantities in inconvenient places. Cobalt is absolutely essential for making batteries, and the Democratic Republic of the Congo has 70 percent of the *world's* cobalt supply. Now, I probably don't have to tell you, but the political situation in the DRC isn't what anyone would call 'stable.' It's not a Republic, it's certainly not democratic, and lots of people die in the process of mining all this cobalt. Oh, and there's lots of children doing the mining.

And rare earth mining is not like striking oil. According to David Abraham, author of *The Elements of Power*, in the Jiangxi rare earth mine in China, workers dig eight-foot holes and pour aluminum sulfate into them to dissolve the sandy clay. Then they haul out the bags of mud and pass the mud through several acid baths, and then they bake what's left in a kiln (powered by coal, obvi). At the end of it all, the metals they're mining for are just 0.2 percent of the shit they take out of the ground. And what do you think they do with that 99.8 percent of mud now contaminated with toxic chemicals? Well, put it right back into the environment, of course!

For a single electric car battery, you have to process 25,000 pounds of brine for the lithium, 30,000 pounds of ore for the cobalt, 5,000 pounds of ore for the nickel, and 25,000 pounds of ore for the copper. All told, you dig up 500,000 pounds of the earth's crust for just one 1,000-pound battery.[23]

And that's just for one car. Now imagine something even bigger . . . like a wind farm. Building a single 100-megawatt wind farm—never mind thousands of them—requires some 30,000 *tons** of iron ore and 50,000 tons of concrete, as well as 900 tons of nonrecyclable plastics for the huge blades. And let's not forget those rare earths like neodymium, praseodymium, dysprosium, and terbium.

And solar is *even* worse. With solar hardware, the tonnage in cement, steel, and glass is 150 percent greater than for wind for the same energy output. And guess what we need to make those complex panels? And guess where we get it? I'll give you a hint . . . it rhymes with shmare mearth mimes.

A better way to think of "renewable" energy like wind and solar is to think of it as "materials energy." A solar-powered car takes free energy from the sun! Hooray! But it does that through materials like silicon for the photovoltaic cells and all that lithium for the battery. And as we know, those materials take energy to get out of the ground—a lot. Renewable energies are no more renewable than a pick-up truck. They utilize the free renewable energy of the sun or the wind, but only by using a lot of materials that break down over time and will always need to be replaced.

And if you're wondering why we call these "rare earths," I'll give you one guess. It's because they're rare. Environmental destruction and oil burning and child mining aside, these materials just can't last forever. And I'm not talking about "forever" like 100 years from now. Forever is pretty fucking soon. Neodymium—essential for the magnets in phones and wind turbines and solar fields and electric vehicles—is expected to run out in 2035.[24]

But let's just ignore that one right now and focus on the elephant in our plan to save the world. If we need to make batteries for every single electric car and batteries so that our solar fields and wind farms can store energy when the sun's not shining and the wind's not blowing, then we're gonna need a whole lotta lithium to get it done. By 2035, our annual demand for lithium is expected to increase by 800 percent from

---

* That's 60 million pounds.

about half a million metric tons to nearly four million. Four million metric tons is 18 percent of allllll the lithium we currently know about and can access on the entire planet (referred to as 'reserves'). If we use 18 percent of what we have every single year . . . how many years is that before it's gone? Before you start screaming at this book about how many lithium deposits we keep finding every year, hold your horses. We're gonna dig into those dynamics a bit later.

But what about good ol' fashioned nuclear fission? That's better, right?

Absolutely! I love nuclear. Let's use so much nuclear. Nuclear has the smallest footprint to energy production ratio of any of the clean energies we have. A piece of uranium has three *million* times the energy of a piece of coal! Take that, stupid fossils!

Looking at the footprint of different energy sources (both how much land the actual plant takes up and how much of the Earth needs to be mined to build it) it takes just 0.3m² per megawatt-hour for a nuclear plant. A regular solar field, by comparison, needs 12.6m² per megawatt-hour.[25] That's 42× more efficient for nuclear.

Which is incredible. And nuclear plants last longer than solar fields. The average nuclear plant lasts for 40 years or more, while the average solar panel (which is mostly nonrecyclable) needs to be replaced every 25–30 years.

AND we might be pretty close to having even better nuclear fission! Bill Gates is working on this new kind of nuclear called Natrium that uses Uranium far more efficiently than the nuclear we know and love (slash hate). I won't go into the details, but it's safer, cheaper, produces wayyy less radioactive waste, and has a much smaller carbon footprint than traditional nuclear! Super cool!

But nuclear isn't a panacea, because nothing is. Guess what we still need to use to mine and refine all that Uranium? Yup . . . oil. And guess what we need to use to make the cement to make the concrete to make the nuclear plant . . . yup. Oil. And nuclear plants use a helluva lotta concrete. I mean, a lot.

But why do we need oil to make cement? Excellent question. First, we gotta get a whole bunch of limestone and crush it up and mix it with silica (sand) and alumina (mined as bauxite, then refined, with oil). Then you burn that in a kiln at 1,400°C (2,550°F) which I don't have to tell you, we need fossil fuels to get that hot. The other thing about making cement is that it doesn't just release CO_2 from all the oil we have to burn; 60 percent of emissions from cement are from the chemical reaction itself, called calcination. The calcination process "liberates" the CO_2 from the limestone and clay, which gives us the calcium we need. This is just science. Limestone + heat = calcium oxide + carbon dioxide. For every ton of cement we make, we get a ton of CO_2.

You know what else we use to make nuclear power plants (and literally any building)? Steel. You know what uses a whole bunch of fossil fuels and emits a whole lot of carbon? Making steel. This involves an even hotter oven, and a bunch of iron and coal. The steel-making process is so energy intensive they literally cannot connect to an electric grid—they'd crash it. So, most steel-making companies just burn coal directly onsite. And once again, we don't just have the CO_2 from powering the kiln, but also the CO_2 as a natural byproduct of mixing the coal and iron. Making one ton of steel produces 1.8 tons of carbon dioxide. Manufacturing steel is responsible for nearly 10 percent of all global carbon emissions—more than all the world's cars!

It's not that these numbers are terrible or insurmountable (though they're not great), it's that even if carbon dioxide wasn't bad for the environment (which it really, really is) we still need fossil fuels to do any of these things.

We can't make any of the energies that are going to save the planet without oil. We can't build nuclear reactors without oil, we can't make electric cars without oil, we can't make deep geothermal plants or wind farms or solar fields or even electric bikes without oil. Even if we keep finding new lithium deposits for the rest of time—even if we invent a whole new kind of battery with a whole new material—we can't dig for that material without oil.

Mining equipment cannot run on electric batteries. Airplanes can't run on electric batteries. Most pick-up trucks can't run on an electric battery. We can't refine the bauxite into aluminum or ship the car across the ocean without oil. We can't drive a wind turbine down the highway without diesel. And no matter how much safer Natrium is, we still can't put nuclear reactors in all of our cars and trucks and boats and construction equipment. And even if we could . . . we'd need oil to do it.

The renewable energy we have right now can't power the processes we use to make renewable energy. Sort of fucked up, huh? Or as Nate Hagens says, renewable energy can power a great civilization . . . just not this one.

But scientists are so smart! I'm sure they'll figure it out!

I know! They really are! Is it possible that we, as a society, could develop an alternative fuel strong enough and dense enough and transportable enough to take the place of oil? Absolutely! Let's do that! I just saw the largest hydrogen-powered aircraft to date was able to fly for ten whole minutes! And it seats NINETEEN people! It's no container ship, but I guess it's something. Except, it won't surprise you to learn we use a ton of oil to make those hydrogen cell engines, and converting the hydrogen into fuel is a very energy intensive process. Not only that, it takes three times as much energy to compress hydrogen into fuel as natural gas. And we lose a lot of energy along the way, basically turning 100 kilowatt-hours of normal energy into 60–80 kilowatt-hours of hydrogen energy.

To make things worse, according to a couple guys at Cornell,[26] creating "blue" hydrogen has a carbon footprint 20 percent greater than using either natural gas or coal directly for heat, or about 60 percent greater than using diesel oil for heat. Right now, we're taking natural gas and converting the methane into hydrogen and capturing the CO_2. But then we're also using natural gas to power the process in addition to the gas we're converting to hydrogen. Doesn't sound great. Just making the hydrogen we already use (mostly to turn into fertilizer) currently emits more than the entire aviation industry combined. The only bonus here is that releasing methane is real, real bad for the planet. Somewhere between twenty-five and eighty times worse than carbon dioxide (depending on your time horizon). So *not* burning it is a good thing. I guess. Except that hydrogen itself, when released into the atmosphere, is *still* way worse than carbon dioxide. Some reports[27] estimate that hydrogen is 11× more warming than good ol' CO_2. Surprise!

Then there is another kind of hydrogen: green hydrogen! Sounds better already! This is where we make hydrogen by putting water through electrolysis (with electricity hopefully supplied by solar, wind, or hydroelectric power) and the water is separated into hydrogen and oxygen. But it's nowhere near commercially viable yet. If we wanted to produce all of today's dedicated hydrogen output using renewable energy, we'd need more electricity than the annual amount generated by the entire EU. That's on top of already doubling the electric grid for all the electric cars we're adding. And that's just for current demand. That doesn't involve a single hydrogen-powered container ship. Oh, and it's big and bulky and hard (and dangerous) to move. The tanks have to be three times larger to power the same sized vessel, and you have to store it right around absolute zero (-252.87°C), and it's super "leaky," so it's always escaping into the atmosphere anyway, aaaaaand it's super-duper explosive.

But I'm not all doom and gloom! This could be promising. Eventually. One day. Maybe? But honestly, we're better off finding some crazy way to directly electrify large-scale transportation then we are exploding this highly flammable, unstable, inefficient chemical.

Just in case you were starting to feel hopeful about Natrium and hydrogen fuel solving all of our problems, I've got one more fun fact to burst any bubbles you may have floating around. Turns out, we don't just burn fossil fuels to operate our heavy machinery and drive our chemical processes and build all of our renewable energy sources. On top of all these indispensable processes we absolutely cannot do without fossil fuels, there are also lots of things we can't *make* without oil.

Think about ALL the plastics we use, from our laundry detergent bottles (and the detergent inside) to our trash bags to our kids' dishes. AND . . . ROADS! We can't make our roads without oil either. The petroleum we use to make roads is a gross, sticky byproduct called bitumen (that we don't use for anything else really) so it's not in addition to our current demand, but we still need barrels of crude to get it. No oil, no asphalt. Petrochemicals derived from oil and natural gas make it possible to manufacture 6,000 different products we use every day, from plastic bags to synthetic rubbers, polishes, waxes, lubricants, motor oils, from adhesive to antihistamines. Aspirin, antibacterials, cough syrups, face creams, ointments, salves . . . they all need petroleum byproducts to exist.

Here is a fun little list from the US Department of Energy, in case you weren't aware we use these hydrocarbon byproducts in pretty much everything.

Adhesive	Credit cards	Lipstick	Safety glasses
Air mattresses	Curtains	Loudspeakers	Shampoo
Ammonia	Dashboards	Lubricants	Shaving cream
Antifreeze	Denture adhesives	Luggage	Shoe polish
Antihistamines	Deodorant	Model cars	Shoes/sandals
Antiseptics	Detergent	Mops	Shower curtains
Artificial limbs	Dice	Motorcycle helmets	Skateboards
Artificial turf	Dishwashing liquid	Movie film	Skis
Asphalt	Dog collars	Nail polish	Soap dishes
Aspirin	Dyes	Noise insulation	Soft contact lenses
Awnings	Electric blankets	Nylon rope	Solar panels
Backpacks	Enamel	Oil filters	Solvents
Balloons	Epoxy paint	Packaging	Spacesuits
Ballpoint pens	Eyeglasses	Paint brushes	Sports car bodies
Bandages	Fan belts	Paint roller	Tennis rackets
Beach umbrellas	Faucet washers	Pajamas	Tents
Boats	Fertilizers	Panty hose	Tires
Cameras	Fishing boots	Parachutes	Tool boxes
Candies and gum	Fishing lures	Perfumes	Tool racks
Candles	Floor wax	Petroleum jelly	Toothbrushes
Car battery cases	Food preservatives	Pharmaceuticals	Toothpaste

Car enamel	Footballs	Pillow filling	Transparent tape
Cassettes	Fuel tanks	Plastic toys	Trash bags
Caulking	Glue	Plastics	Tubing
CDs	Glycerin	Plywood adhesive	TV cabinets
Cell phones	Golf balls	Propane	Umbrellas
Clothes	Guitar strings	Purses	Upholstery
Clothesline	Hair coloring	Putty	Vaporizers
Coffee makers	Hair curlers	Refrigerants	Vinyl flooring
Combs	Hand lotion	Refrigerator linings	Vitamin capsules
Comp. keyboards	Heart valves	Roller skate wheels	Water pipes
Computer monitors	House paint	Roofing	Wind turbines
Cortisone	Insecticide	Rubber cement	Yarn
Crayons	Ink	Rubbing alcohol	

According to the US Energy Information Administration, or EIA, 7 percent of fossil fuel hydrocarbons we use in the US are to make STUFF. Not to burn for energy.[28] And there is no other material on the planet that can replace it. The hydrocarbon bonds we find in fossil fuels are super special. They're unusually strong; that's where the energy is stored. But if we don't release that energy by burning it, it makes a nearly indestructible product—hence why it takes that plastic water bottle so long to degrade when you throw it out your car window.

There are some fun ideas where we start taking all the plastic trash we've made and somehow turn it into roads in the future. Or other ideas where we replace a ton of plastic packaging with hemp-based alternatives. These are things we should be working on (and some people are working on), but right now, we just take the byproducts from oil refining to make these things because it's so much cheaper to do that than it is to recycle old plastics.

In fact, virgin plastic is so much cheaper, some Chinese factories are mixing virgin plastic into their recycled plastic so they can pass it off as 100 percent recycled because of how many companies are "going green." The cost of virgin PET (the most common plastic we use for drink bottles) was $500–600 a ton in 2020, while recycled PET flakes cost $1,000 a ton.

Which is insane. What's even more insane is the demand for recycled plastics (between companies choosing to go green and new regulations requiring companies do it) has driven up the price even further. The price of recycled plastic has increased nearly 100 percent since I wrote that last paragraph and is now $1,916 per metric ton![†] Now, markets are complex, so this can end up being a sort of good thing, because higher prices make it more attractive for people to get into the business of plastic recycling, and if we have more supply that should drive prices back down, but fuck me. We really need that plastic.

Okay, enough about how "plastics make it possible."[‡]

On top of all the products we couldn't imagine living without, there is one we indisputably can't live without—fertilizer. I can't stress this one enough: our entire modern agriculture industry is completely reliant on nitrogen-based fertilizer. Fifty percent of the world's food production couldn't exist without it. And that's only increasing as countries continue to grow and industrialize. Where does that nitrogen come from? Natural gas. We take that methane and split out the hydrogen and use that hydrogen to make ammonia. And we use fossil fuels to power the entire process.

Since we've spent the last few millennia destroying the rich, organic, life-supporting soils by over-farming them (and we stopped using human and animal waste to complete the cycle and naturally fertilize), we have to keep pumping more and more fertilizer into the ground to get crops to grow every season. Pumping those crops full of ancient nitrogen certainly helped us explode the population in the past couple centuries. Otherwise, we would have been limited to what the ground could actually regenerate. That's no fun. Here's a fun chart showing how much the world population wouldn't have grown without it: [29]

---

† A metric ton, or 'tonne' weighs 2,204 lbs., or exactly 1,000 kg. The imperial unit 'ton' weighs 2,240 lbs., or 1,016 kg. I'll be switching back and forth throughout the book, but just know they're pretty much the same.

‡ In case you weren't old enough, the plastics industry came up with their own advertising campaign in the nineties juuuuust in case we weren't already using enough plastics. "Plastics make it possible" was, and still is, their slogan.

World population with and without synthetic nitrogen fertilizers

Estimates of the global population reliant on synthetic nitrogenous fertilizers, produced through the Haber-Bosch process for food production.

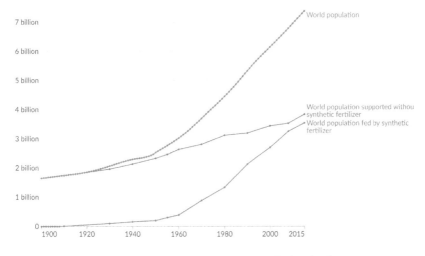

Data source: OurWorldInData.org/how-many-people-does-synthetic-fertilizer-feed | CC BY

At this point? We're basically just converting fossil fuels into food. Think about that: OUR FOOD is a fossil fuel product. Fifty percent of the molecules in your body came from natural gas that we converted into ammonia that allowed those crops to grow. And the cherry on top? The process we use to convert the natural gas into ammonia-based fertilizer (known as the Haber-Bosch process) releases a shit-ton of carbon. Because of course it does.

The Haber-Bosch reaction runs at high pressures at temperatures around 500°C, sucking up about 1 percent of the world's total energy production, and it belched up about 451 million tons of CO_2 in 2010.[30] That accounts for roughly 1 percent of global annual CO_2 emissions, more than any other industrial chemical-making reaction. And that's on top of allllll the natural gas we need to convert to fertilizer in the first place.

But without it, half the global population would starve to death, so I guess, carry on.

When you think about all the stuff we absolutely couldn't live without that comes from these incredibly valuable, irreplaceable, miracle substances that are fossil fuels, it starts to seem pretty insane that we just put it in our cars and explode it to get places faster. Future generations are probably going to look back at how much of it we just set on fire and be like, "What the fuck were they thinking???"

The Perpetual Motion Machine

**Well first of all, tell me: Is there some society
you know that doesn't run on greed?**

—Milton Friedman

Now that you're feeling hopeful and energized and ready to take on the day, let's talk about the other system we have that runs on fossil fuels: CAPITALISM. Our great, shiny economic philosophy that's become more akin to a religion in America. The Incredible Everything Machine is brought to you by capitalism. Capitalism is what increased living standards across the whole planet. It's what brought seven billion of us out of poverty. It's what has allowed eight billion of us to exist. It brought us globalization and specialization and the division of labor and all the cheapest things everywhere all the time!

But what is it, exactly? Capitalism is a system where trade and industry are controlled by private owners for profit, as opposed to say, socialism, where industry is community-owned, and any "profits" or benefits are shared among everyone. In a capitalist system, private citizens own businesses and sell their products for a profit, and the prices that those businesses sell things for are set by the market rather than by the government. I can charge whatever I want for my radishes, and as long as someone is willing to pay it, we're all good. The government can't tell me my prices are too high, and they can't force me to sell my radishes for any less. You probably already knew all that, but this book is meant as a primer, so we're stripping everything down to its birthday suit. The best part about capitalism is that the competition at the core of it makes it a super-efficient machine: everyone figuring out the best way to make money (so they can eat and buy the things they want) is the most efficient way to divvy up the resources of a society.

So far this all sounds pretty good. People work hard so they can make money so they can eat (and buy things). And the more money people make, the more things they can buy—like food, water, electricity, houses. But built into capitalism, the assumption that keeps the whole thing humming, is the profit motivator. It's what (in theory) drives everybody to do everything in the world we've created. People want to make more money. So they will always sell their goods for the highest price possible. And the incentive to get more money inspires people to invent cool new things so they can be the ones to make extra money. And people making extra money and buying extra things means the economy is growing. And growth is good! It means you can get richer! It means everyone can get richer! From what I've been told about capitalism since the day I was born, it's working great. The more people buy, the more money circulates around the world, the better off *everyone* is. We can finally get those last billion people out of poverty. Socialism failed, communism failed, capitalism is THE SYSTEM.

There's just one tinyyyyyyy problem. The capitalism we have kinda needs the economy to keep growing all the time—or we immediately descend into crisis.

Grow, baby, grow

You've probably heard me mention GDP a few times and thought, yes, of course, I know what GDP is. But just in case, GDP, or Gross Domestic Product, is used to talk about the size of national economies. It's really how we've come to define what an economy is. It's capitalism's favorite number. GDP is the total amount of all the things being bought and sold in a nation. It includes private investments as well as government spending. It includes things we produce here but sell to people in other countries. It's basically a measure of all the money that changes hands within a country. It's all the money being made by everybody.

It's also the way we've decided, for some asinine reason, to measure the health of our economies and by extension, the health of our countries. It's become a universal measure of success without a single competing metric in our global economy. It's the only number anybody cares about. All any president talks about is the economy growing. If the economy isn't growing, he'll get voted out of office. We have teams of economists making massive global and domestic policy decisions all based on making sure this single number never, and I mean NEVER, goes below zero. And we really like to keep it above 2 percent in fully industrialized economies.

But the crazy part is, while most of us have lived our entire lives under the all-powerful reign of GDP, it's actually pretty recent. The concept of GDP wasn't even invented until 1934 in the wake of the Great Depression, and it wasn't until after World War II, with the Bretton Woods agreement in 1944, that it became the global standard of measurement it is today.

Why asinine, you say? Well, as good as GDP is at telling us when people are buying stuff, I shouldn't have to tell you that not every single part of the human experience—or even the economy—is connected to money changing hands.

There was a time before GDP when economic activity was thought of in a more holistic way. It was thought of as both physical materials (stuff you buy) and energy flows (things you do). In this framework, raising your children or taking care of your aging parents were both flows of energy that were useful for an economy, for a nation. But since neither of those things are paid work, they couldn't be included in the all-powerful GDP.

In addition to the incredibly valuable unpaid labor of child-rearing, GDP also doesn't account for non-monetary costs, like environmental damage. It doesn't account for happiness or wellbeing. Okay, sure, some people are getting richer . . . but how many people are happier?

GDP also does this hilarious thing where bad things can make it bigger. Let's say I smoked for twenty-five years, and then I got cancer and went to the hospital and had to pay for chemo and radiation and whatever other treatments. Those cigarettes were added to GDP, obviously. But those cancer treatments *also* get added to GDP. Everybody in the country could start smoking and start getting cancer and economists would be like, Hooray! GDP is growing again! But we can see quite clearly that is not good for anyone.

Or how about oil spills? When BP spills, say, I don't know, 3.19 billion barrels of oil into the Gulf of Mexico covering more than 44,000 square miles, that is clearly not good for the environment. Or for the people who live near the Gulf of Mexico, or literally anyone or anything. But you know what that oil spill did? It *increased* GDP. Because with an oil spill comes clean-up missions. BP spent $15 billion on clean-up and another $20 billion in economic damage payments. This isn't growth—more people aren't buying and selling things and living better lives. It's just bloat.

Now, GDP was real useful in the aftermath of the Great Depression so people could figure out if the economy was actually getting back on track or not. Are businesses making things and are people buying them? But the dude who invented it, Simon Kuznets, never intended it to become the all-encompassing determiner of national and global policymaking for the rest of time. As he said himself, "The welfare of a nation can scarcely be inferred from a measure of national income." Or as Oxford economist Diane Coyle says it better, "[GDP] tells you how valuable trees are when you cut them down and turn them into fences or benches, but not what they're worth when left standing."

We'll talk a lot more about GDP and wellbeing later. But for now, we know that GDP is a flawed measure of a nation's health or wealth or happiness. But we also know that GDP growing in very poor countries is usually a very good thing. It means more people are being lifted out of abject poverty and starvation. It means more people with electricity and education and clean water. Cool cool cool.

And in more advanced economies, when GDP is growing, that's still great! It means people in your country are buying more things and selling more things and making more money, which means they can eat and go to the movies and buy even more things! (Are you seeing a pattern here?)

So pretty much everyone has accepted the narrative that GDP should always be growing. Poor countries want their GDP to grow because it means more of their people can buy food and heat their homes. Rich countries want their GDPs to grow because . . . who doesn't love money? Every country in the world wants their GDP to grow every year. Which means the economy of every country gets bigger every year. Which means, obviously, the global economy gets bigger every year.

Here's a fun chart showing how good we are at making it bigger:

Global gross domestic product (GDP) at current prices from 1985 to 2028

Total output of the world economy, adjusted for inflation and expressed in billion USD

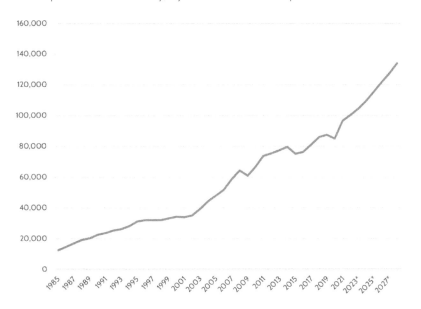

Data Source: IMF World Economic Outlook Database

Look at us go!

Or, actually, let's look at a chart going back a couple of millennia, just for fun:

Global GDP over the last two millennia

Total output of the world economy, adjusted for inflation and expressed in 2011 $

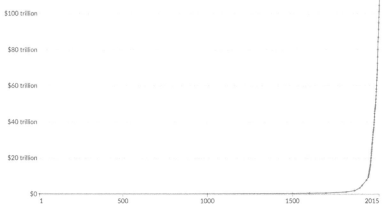

Data Source: Our World in Data

Now that's more like it!

Every country in the world has financial levers and mechanisms they can pull to help make sure this line keeps on climbing. The EU just started including cocaine and prostitution in their GDPs so they could boost their growth rate, which I'm not mad at, I guess. Those are both legit businesses.

The thing is—and I can't stress this enough—we don't just *want* it to grow. We *need* it to grow.

I've often wondered about this. Why does every business have to grow? Why can't we just exist? How come I can't just open my restaurant down the street, and it makes enough money to feed my family, and that's that? I don't have to expand into another location, or start a franchise, or squeeze in a few more tables. Sure, some people do that because they want to be the most successful person in the world or because they're greedy (or because society has taught us that money is the only thing with value, so we all try to accumulate as much of it as possible). But why can't I just run my restaurant and it feeds my family, forever? Well, you sort of can. So long as you keep raising prices in line with inflation. So technically your business is growing in terms of revenue, even if you have the same number of tables in there.

But again, why?*

Well, part of it is that when the economy contracts (when GDP goes down), we have a recession. And when there is a recession (or a depression), people lose their jobs. And then they don't have money, so they stop buying stuff, so other people have to close their businesses because they aren't selling enough stuff. And then those people can't buy things either, and it snowballs pretty fast. Since those things are bad, we really want GDP to keep growing. Hooray, everyone has jobs! Now, some very complex economic engines are at play here that admittedly, I don't fully understand. Or maybe I do, but they're just dumber than they seem. For one, massive corporations lay people off when they expect a recession (which is happening right now), which can actually end up contributing to the recession, which is pretty stupid if you ask me. The other thing is that economists have their heads so far up their asses when it comes to GDP that unemployment being "too low" is a real thing. When unemployment is too low, it can lead to wage inflation (people being paid "too much"), which causes businesses to struggle and also leads to "reduced productivity," which again, hurts profits. And since profits are all economists care about (since GDP doesn't measure things like wellbeing), sometimes they actually *want* more people to lose their jobs so we can save the precious economy—and keep that GDP growing. Call me crazy, but wouldn't higher wages mean more people could buy more things, which would be, I don't know, good for the economy???

I don't think I have to remind you that in the early days of the pandemic, some politicians were suggesting we all let our grandparents die in order to save the economy.

---

* There is a lot of economic argument around the growth imperative, but I am not an economist. So I'm not gonna dig too deep into all the different retorts people will have to the statement that our current system requires growth. Sure, a zero-growth version of capitalism is theoretically possible. But that's not what we've got, and we do everything in our power to make sure we keep growing.

No one reached out to me as a senior citizen and said, 'Are you willing to take a chance on your survival in exchange for keeping the America that all America loves for your children and grand-children?' And if that's the exchange, I'm all in. [. . .] There are lots of grandparents out there like me.[31]

—Texas Lieutenant Governor Dan Patrick

Anat Shenker-Osorio
@anatosaurus · Follow

Some of You Have to Die. Wasn't where I saw the GOP 2020 slogan going but I appreciate the candor.

12:53 AM · Mar 24, 2020

13.5K   Reply   Copy link

Read 125 replies

But why?

At the core of our desperate need for endless growth is: debt.

The vast majority of businesses (and economies) in the entire world run on debt. You take out a loan at a certain interest rate. And you promise to pay back that person, plus whatever interest rate you agreed on. But that interest rate means growth is required. If you're not growing, you'll never be able to pay them back. If you don't grow enough to cover the interest, then eventually the debt gets bigger and bigger and then, like Greece in 2008 (like a lot of things in 2008), things collapse.

Like a three-year-old in an endless line of inane questioning, I'll ask once again . . . But why?

We have too many negative pigs

Paper money (or its digital equivalent) isn't real; it represents a debt. Real wealth is represented by things. Like a pig for example. You can have two pigs and that's twice as good as having one pig. But debt isn't a real thing that can exist . . . you can't have negative pigs. Think about negative pigs like unicorns. Let's say I owed you some money, but I said I'd pay you back with a pig. You might think "Okay, Taylor is definitely gonna give me that pig at some point." But you probably wouldn't take an "IOU one unicorn" as payment—because unicorns don't fucking exist.

Except, our entire economic system is based on people owing other people unicorns hoping they turn into pigs. When you walk into the bank to take out a loan, you probably think they are lending you money they already have. But they aren't. The banks only have about ten percent of the money they use in actual money. So they just keep creating fake money every time they loan out money to people or businesses. They just keep creating fictional pigs, betting that next year, there will be more pigs. Maybe those pigs exist, or maybe they're unicorns and they never will. And as our economy has become more and more fueled by debt since the 1970s, we're constantly "creating" wealth that doesn't exist. There are no physical pigs to back it up. We have too many unicorns, too many negative pigs. And it isn't just the banks or the US government— every country in the entire world is doing this, to varying degrees of financial risk and instability.

So, when someone can't make payments on their debt—when they don't make as many pigs as they thought they would—the interest compounds, turning into principle, increasing the overall debt. Instead of owing you two pigs, now I owe you three, even though nothing really changed. Another negative pig was just created out of thin air. Each day or month or year that passes, it grows larger and larger. And I have to grow even more if I ever want to pay you back. Of course, this isn't real money or real pigs or anything. It's virtual, theoretical wealth. And when it gets so big that the borrower can't make the interest payments at all, and the debt is first defaulted on, then canceled (again, as in the housing crisis of 2008), that virtual wealth is destroyed. Most of it never really existed in the first place. But it did exist in the sense that when a bank loses billions of dollars, real, actual people lose their retirement funds, their life savings, their houses, and their pigs.

But how can people just invent pigs? Why do we let this happen? Couldn't we just *not* do that?

Well, this whole debt idea really got rolling with a British dude named Maynard Keynes in the 1930s—an exceptionally tumultuous time in the world that included the Great Depression sandwiched between two World Wars. So after the Depression hit, Keynes came up with the idea that we needed more government intervention and more stability and MORE DEBT. This made a lot of sense at the time. The United States probably wouldn't have come out the other side of the Depression if it weren't for all the government-funded social programs of the New Deal, which were funded by debt.

Except back when Keynes came up with these ideas, we still required paper money to be backed by actual wealth (we used gold instead of pigs). We had the gold standard. And all the countries in the world abided by it. You knew that if you wanted to trade in your $10 bill to the US government, they would give you $10 worth of gold. But one tricky thing happened during World War II: the United States kind of took all the gold out of Europe. We demanded payment for everything in gold. So, after the war—since no other country had enough gold to back up their own currencies—the US dollar was basically it. Here's another passage from Peter Zeihan's endlessly entertaining and informative book:

> The US dollar wasn't just the only reasonable medium of exchange in the entire Western Hemisphere: it had sucked the very metal out of Europe that would have enabled a long-term currency competitor anywhere in the Eastern Hemisphere. If anything, this is truer than it sounds. After all, the metals-backed currencies of Europe were the culmination of *all* human civilizations of *all* eras stripping the entire planet of precious metals since *before* the dawn of recorded history. And now it was in Fort Knox.

By 1947, the US had 70 percent of the gold in the entire world. Countries around the world were ravaged from war and the Depression. Inflation was over 1,000 percent in some places. European currencies were nearly worthless. No one had any money (except America), and life was hard and shitty. But boy oh boy, was Keynes on the scene telling everybody what needed to be done. He was like pull out those credit cards guys, we're going on a shopping spree!

Get in loser, we're going shopping.

We got the Bretton Woods conference, which set new rules for an international monetary system; we created the International Monetary Fund (IMF) to help out struggling economies; and we birthed the concept of globalization to stabilize the world order. Keynes was like, "America will lend you losers some of these precious dollars you can't get anywhere else. All you have to do is get back to work in those factories making stuff. And then promise to use some of those dollars to buy a bunch of awesome American stuff." Everybody's in debt to buy a bunch more stuff, but everybody's economy is growing. And it worked. It worked like gangbusters. We got all wrapped up in GDP and the excitement and prosperity of endless growth and never looked back.

But even with all this deficit spending, it's important to note we *still* weren't inventing pigs. All that money still had to be tied to gold. So we had to keep finding gold in order to keep issuing debt. Which was a bummer because we were all super excited about buying and building more stuff after the war. Like SO much stuff. Like all of suburban America and the interstate highway system and skyscrapers and ... pretty much everything we take for granted today as the American way of life.

So once there wasn't enough gold to let us continue to grow as fast as we wanted to (because we had so much oil to burn!), it didn't take long for the US government to say, "Look, we're not gonna back the dollar with gold anymore, you're just gonna have to trust us; we're good for it." This happened in the 1970s—just fifty fucking years ago. So now, you could basically just invent money while saying, "We're good for it," which everybody started doing.

Whereas before growth was limited to the physical world—or how many pigs actually existed—there was no longer any practical cap on the capital available to do whatever you wanted. Our new economic measures (GDP) just looked at how much paper money was changing hands. And since we could print as much paper money as we wanted, and paper money magically expands over time as a function of debt and interest rates, we could grow and grow and grow without those pesky limits of physical boundaries.

Today, global debt is 247 percent of global GDP. Think about how much money you make every year. Now imagine you owed 2.5× that amount in credit card debt. How long would it take you to pay that off? In order to pay out all of our global debt claims without some great collapse (or lots of tiny collapses everywhere), we need to more than double the size of the global economy—and that's just to catch up to what we've already bought and built and used and thrown away.

Super fun, right!

So debt in general is a claim on future wealth, future productivity, future growth, future pigs . . . so what? No big deal, right? If we can just keep printing money that doesn't have to be backed by any physical resources, then we can just keep growing forever . . . right? Money is money? Right???

Wrong.

Money isn't money. Money is stuff. And all the stuff we make and buy and use takes energy. As we've already discussed, our entire economy runs on energy. To put it bluntly, the economy IS energy. There is no wealth, no iPhones, no clothes, no grocery stores without energy. You can't even throw something away without using energy.

So when we create debt to be repaid in the future, we're essentially stealing energy from future humans. Because energy is needed to create the stuff to create the money that will pay back that debt in an ever-growing system. I can't grow my restaurant to pay back that loan without using gas to light the stoves and electricity to keep the lights on and the fridges running, and all the diesel that's used to transport all the food I'm going to cook and serve that gets delivered from every corner of the country (and planet). All of the growth we need to avoid collapse is based on energy usage. I have to keep growing and keep using more energy because the interest rate on my business loan is six percent. If I don't grow, I'll lose my restaurant!

And if the United States doesn't grow, we won't be able to pay the interest on the trillions of dollars of debt we've accrued. In 2022, the United States paid nearly half a trillion dollars just in interest. Yikes.

When seen on a global scale, to keep our happy capitalist system humming, we have to keep the global economic machine growing by around three percent a year . . . or we plunge into crisis. That means we don't just have to double the size of the economy once. We have to double the size of the economy every twenty years just to stay afloat. At 3 percent growth it means an economy twenty times bigger in the next hundred years. An economy that has already been largely funded by claims on future energy. An economy where growth correlates to near perfect growth in energy usage and where we have no current viable alternative to replace our main input. Knowing everything you know about how many people are alive and how much stuff we eat and buy and throw away . . . and how much energy we have left to burn . . . does that sound reasonable to you?

Random, ravage, rust & rot

In case that still sounds reasonable to you, let's talk about entropy for a second. Entropy can be a confusing concept, but it's critical to understanding how the entire world functions and why our current models are both unsustainable and insane.

Everything in the world has energy inside of it. A piece of wood, a cell phone, a barrel of oil. Entropy is a measure of how much energy in a thing is *un*available for use. So oil is low entropy, because you can turn that whole sucker into usable energy right now. But your cell phone is high entropy, because, while it does a lot of cool shit, you can't actually turn it back into anything useful when it breaks. You just throw it away and buy a new one.

The First Law of Thermodynamics says that energy can neither be created nor destroyed; the system we live in is a closed system. Everything that can exist already exists in one form or another. In order to make an iPhone, you have to find all those materials and put them together. And whether you recycle the iPhone or throw it away or burn it, those materials will eventually turn into some form of higher-entropy energy (some form of waste). They never go anywhere or disappear; they just change form.

The Second Law of Thermodynamics is the law of "random, ravage, rust, and rot." It says that energy only moves in one direction. Things go from low entropy to high entropy. Heat will always pass from a warmer thing to a colder thing. There is no such thing as a perpetual motion machine because we will always have to put in some external work (or energy) for it to move in the other direction (or to turn your cell phone back into useful energy). Think about it like your house. Your house gets messier and messier, and the only way to reverse that process is to expend energy while cleaning it: to put things back into a state of order. So entropy can also be thought of as things always going from a state with "low chaos" to "high chaos." Things will always, always, always "get messier" and break down.

As the famed economist Nicholas Georgescu-Roegen wrote in his 1976 collection of essays, *Energy and Economic Myths*:

> Were it not for [the law of entropy], we could use the energy of a piece of coal over and over again, by transforming it into heat, the heat into work, and the work back into heat. Also, engines, homes and even living organisms [...] would never wear out. There would be no economic difference between material goods and Ricardian land.[32†]

---

† Ricardian land is based on classical economist David Ricardo's theory that land is free and never degrades, and so the rent you can earn on a piece of land is essentially boundless.

This indisputable law of physics tells us why we can't just grow money forever and ever, even though our entire system (and debt and interest in general) is based on the idea that we can.

Let's say you were gonna lend out something you own, some physical capital of yours. Maybe some pigs, or your minivan. Those things would deteriorate over time. Your pig would get older; your van would have 30,000 more miles on it. And when your friend returned them to you, they would need to return a different pig of the same age or a van in the original condition. Otherwise, the thing they were borrowing would be worth less money than when you lent it. And you can argue that's what the interest is for: it covers the cost of your deteriorating asset. But for some reason, we don't apply the same logic to money. If you lent your friend $10,000 instead of a van, the $10,000 would still be the same. The money that represented the value of those things is treated as eternally perfect. It never depreciates like a pig or a car. It does the opposite: it *appreciates*. And the interest created from lending out your capital doesn't depreciate either. If they're paying you interest on that $10,000, and you earned an extra $1,000, the new money would make even more money because that's how compound interest works.

As radiochemist Frederick Soddy said back in 1926, "The ruling passion of the age is to convert wealth into debt in order to derive a permanent future income from it—to convert wealth that perishes into debt that endures, debt that does not rot, costs nothing to maintain, and brings in perennial interest."

But if we know that every physical thing on Earth must deteriorate, and therefore depreciate in value, why does money do the opposite? How is that possible when we know money has to represent a physical thing in the world? Great question.

Entropy tells us why an infinitely growing economy doesn't work (and why endlessly compounding interest on debt makes no sense). Because you can turn a single pile of money you have into a continual stream of income, forever. This would be awesome if it were actually possible. But it's not. Perpetual wealth creation from a single stock of non-depreciating capital is literally impossible so long as the laws of thermodynamics hold true. Because that money has to represent a physical thing in the world. And physical things in the world are all subject to the law of ravage, random, rust, and rot.

Infinite Growth in a Finite System

**The market's compulsion to grow outcompetes
any alternate path of wisdom or constraint.**

—Nate Hagens

In case that *still* sounds normal to you, let me offer yet another reason why this whole system is utterly insane.

Capitalism means we keep growing. We keep inventing money, and that money buys *things*. You buy cars and houses and iPhones and new shoes. Businesses build factories and governments build highways. And where do those things, those physical materials, come from? The Earth. A finite ball of resources. It is *finite*. As in, it has a hard limit on the amount of stuff we can take out of it. So whether or not "endless growth capitalism" can last 20 more years or 200 more years isn't really relevant. The point is that it has an end date.

Infinite growth simply isn't possible on a finite planet.

I want you to imagine a square. Actually, don't imagine a square, I'll just show you one. Here, look at this square:

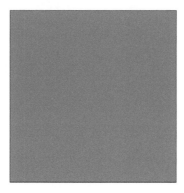

This square represents our economy. Inside the square is every product made and sold and all the money exchanged. The way our economy is set up and thought about is known as the circular flow model: People work for businesses that pay them money, then they use that money to buy stuff from other businesses so those businesses can pay someone else money, who then buys more stuff. In this model, the "stuff" that makes up the things you buy is just represented by more money. And money, as we now know (according to economists) is infinite. According to economists, this cycle can go on forever.

Here is a visualization of the circular flow model:

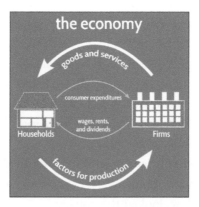

Notice how the line that says "factors for production" just comes out of nowhere . . . or like it comes out of your house? Obviously, that's incorrect. I don't have a rare earth mine in my living room. There has to be an INPUT of physical goods to MAKE the goods.

To make a more accurate depiction of how the world functions, let's draw a circle around the square of the economy. It should look like this:

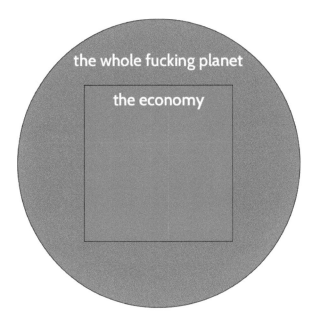

The circle is the environment, our ecosystem, our planet. Our economy (and us and everything we do, eat, buy, and throw away) exists within the boundaries of the ecosystem. Input into the economy comes from the environment and output goes back as waste—whether trash we throw in a landfill or carbon dioxide in the atmosphere or literal toxic waste that corporations dump into our rivers and oceans. Nothing in our economy is made without taking some raw material from the circle, and nothing is made without putting some other form of waste back into the circle: energy can neither be created nor destroyed.

Turn the page to see what your final picture should look like:

Fun, right! It's a closed system. The economy doesn't exist separate from the ecosystem. The waste side isn't separate from the raw materials side. It is one system, fully dependent on—fully encompassed by—the other. And there is nowhere else for waste to go except back into the system. Because that is what a closed system is.

You may have heard of "closed-loop manufacturing." The idea behind closed-loop manufacturing is that we completely reuse the materials to make a product once the original product dies. Since we live in a finite world, and things are always turning from low entropy to high entropy, and nothing can be created or destroyed, we have to consider how much materials we use to make everything we make. This is known as entropic overhead. Entropic overhead is a measure of how good we are at maintaining something we've made (your cell phone or your car) or how good we are at reusing the pieces we used to build it.[33]

For the record, we are very, very bad at this.

Think about an hourglass. When you flip it over, it counts the time for two minutes or three minutes or however much sand you have in there. And then, when it's done, you can turn it back over and it will do the same job all over again! There is zero entropic overhead. The product just keeps doing its job forever (though you could argue the sand slowly erodes and therefore falls a little faster and so it stops keeping very good time, but let's act like economists and ignore that for now). Now imagine, instead of this perfectly circular model, after the top of the hourglass empties out, you just smash it, buy another hourglass, and then maybe try to recycle the old one if you're feeling virtuous (or if the market for recycled hourglass parts is offering a positive ROI).

That's what we're doing with most of the products in our entire economy. We buy them, we use them for a minute, and we throw them away—all the raw materials in there, whether it's the petroleum-based fabric from your umbrella or all that lithium in your car battery or the plastic in your Ziploc bag—we just make it and toss it. So that is 100 percent entropic overhead. Not great. I mean, I guess this would be fine if our planet wasn't a finite ball of resources, except it is finite. We've only got the one ball.

Now for the even crazier part. The craziest part about all of this isn't how much debt we're in, or that debt doesn't even make sense, or how much oil we use, or that it's definitely going to run out (it is), or how many perfectly good things we throw away every day, or how we're going to run out of those too . . . but that modern economists don't even account for *any of it* in our economy. The craziest part is that no one is even talking about this, and if you ask an economist, they'll tell you you're wrong. They'll tell you our economic system can continue to grow infinitely while they worship a little shrine to Milton Friedman in their closet.

Because economics somehow along the way got completely detached from the ecological system our entire lives depend on.

In our system, smashing the hourglass is actually a good thing.

Because it means you had to buy a new hourglass, which means you spent more money, and GDP got bigger. Our current economic system doesn't account for the material input to make an hourglass, or the trash created by throwing out the old hourglass, or the energy it takes to make a new hourglass, or the pollution created by burning that energy to produce all those new hourglasses. The materials to make the thing were never even a part of the equation. Our entire monetary system is essentially betting that the Earth itself continues to get bigger and bigger every year—that we can make an infinite number of hourglasses and iPhones while just tossing the old ones in a landfill.

The problem is that some super cool dudes like 200 years ago decided that our entire economy is driven by just four inputs: labor, capital, land, and entrepreneurship. Or put more simply: work, money, land, and ideas. No materials, no energy. Entropy shmentropy.

According to our economic models, money creates more money and productivity increases eternally.

The "nature" in our model—from which every single product is made and every machine is powered—is completely absent. It's just represented by money. None of the equations that we use to make assumptions about what's good for our economy (or us or our planet) account for the physical goods that have to come *from* the planet to make any of the things. And most importantly—they don't account for the massive amount of energy we're sucking out of the system that replaces human labor, or for the waste that burning that energy puts right back into it.

To be clear, this is still what economics textbooks are teaching. Seriously.

Our entire modern economic model is just the little blue square. There isn't a green circle around it at all. It's a perpetual fucking motion machine. Hourglasses and iPhones materialize out of thin air. Every year that blue square gets bigger and bigger and bigger, and that green circle stays the exact same size. And nearly every economist on the planet tells us we can absolutely continue to grow the blue square forever. And not only that we *can*. But that we *should* and we *must*.

At this point, you're hopefully wondering, but HOW? How could they be so stupid? I mean, it's so obvious!

I knowwwww.

To be fair to economists, the Perpetual Motion Machine wasn't always the preferred theory. The very first economic theorists in mid-18th-century France, known as the physiocrats, had a much more holistic view of things. They explained economics in line with natural law. They saw Mother Earth as the source of all value (because, obvi). The regeneration of plants and animals was the basis on which all wealth creation was understood. This was the first concept of circular flow: Farmers grow food, and then we buy food, and then the farmer uses that money to buy things from us, and then Mother Nature grows more food the next season. For the physiocrats, money was sterile; it was just a thing that represented the actual things. And they completely rejected the idea that money could "reproduce" through interest because it didn't fit their very logical paradigm of value coming from the Earth. The critical piece to note in this circular-flow model is the ability of Mother Nature to regenerate the plants and animals every year.

Unfortunately, we sort of have Adam Smith to blame for what happened next. And Karl Marx. And the "Ricardian land" guy, David Ricardo. While capitalism came into being in the 16th century, Adam Smith is often thought of as its father. Because he wrote a book in 1776 called *An Inquiry into the Nature and Causes of the Wealth of Nations*. This book has basically become the Bible for free-market capitalists.* We're gonna talk lots more about how wrong those people are later. But the more important piece of *Wealth of Nations* to talk about now is the labor theory of wealth.

According to Smith, the value of a thing isn't inherent, or based on what the thing is made out of, but derives from the amount of labor that thing could command from others. If I have a rocking chair, maybe I can trade it for five scarves. And if those five scarves each took five hours to

---

* More accurately, a painful misinterpretation of this book from people who've clearly never read it has become the Bible for free-market capitalists.

make, my rocking chair is worth twenty-five hours of labor. We don't have to get into too much detail, but both Marx and Ricardo viewed the labor theory of wealth from a slightly different angle. For them, the value in a thing is based on how many hours it took to make the thing itself: how many hours it took you to build the rocking chair rather than how many hours it took to knit the scarves you would trade for that chair.

Remember, this was the 18th and early 19th centuries. Before I could even start thinking about knitting a scarf, I had to raise my own sheep and then make my own yarn on a loom that I built out of a tree I cut down with my bare hands (and an axe). Or ideally, in capitalism, one guy raised the sheep, one guy made the axe, another guy built the loom, and then I could spend five hours focusing on the knitting. So many hours of labor went into that scarf that when you buy it (instead of making it yourself) you're buying all that labor. The more money you have, the more other-people's-labor you can buy. The same is true for literally any product or service you can think of. Which, they're not *wrong*, it's just not the whole picture. Whether you're looking at how much labor the thing demands or how much labor it took to make it, they were over-looking a couple key points.

Obviously, the human labor is a very big part of the value of the thing. But so are the sheep. And so is the tree you needed to cut down to make the loom. And the wood and steel for that axe. And if our loom is steam-powered, you can add the wood or coal or oil to burn in there too. All of these things are inputs. Where we got confused—and how all of these inputs got taken out of the equation—is that the materials to make a scarf weren't really a big concern back in Smith's day. All of those scarves and rocking chairs were still being made by hand. They hadn't struck oil yet. Factories weren't electrified yet. All the labor had to be human labor because it's pretty much the only labor they had in 1776. And they certainly weren't worried about running out of sheep or trees—two things that can infinitely reproduce through the bounty of our dear Mother Nature.

But once we started inventing machines that made humans more productive, we sort of lost the plot. Shit went absolutely nuts. Suddenly, with a spinning jenny, you could make scarves eight times as fast, which meant eight times less human labor. The materials stayed the same, but the human labor input went way down. And when that shot productivity and profits way up, these guys—who were literally inventing the concept of economics on the fly—felt like they had hit the jackpot. There would always be sheep and wood and whatever else. Substituting "money" in our equations for the physical goods made just as much sense. Suck it, physiocrats. The labor and land and capital and ideas were enough to describe how the economy was working.

And for the next two hundred years, that's pretty much all these white dudes talked about: labor and capital and innovation and ownership of the means of production. Workers and money and ideas and who profits from it all. It's still all they're talking about. When Karl Marx dropped those two bangin' critiques of capitalism—*Das Kapital* and *The Communist Manifesto*—everyone got all worked up over which white guy was right so they could talk about it some more. Neoclassical economic theory[†] came out around the same time, and we started to dig into rational consumer behavior and firms existing solely to maximize profits (because CAPITALISM!) and different tools we can use to maintain market equilibrium. All these dudes became obsessed with opining on how one thing might shift the levers of the market one way or another. If people were wearing fewer scarves, what would happen to the price of scarves? What would happen to wool production? They couldn't shut up about supply and demand and keeping the market stabilized at all costs, and if you do that, this will happen in the market . . . and how can everyone get richer while we do it?

And that's how we ended up with an overarching economic theory that didn't include any of the actual *stuff*. It never occurred to anyone to ask, *Where exactly are we getting the wool from? What will happen if the sheep go extinct?*

And this whole time while a bunch of bougie white dudes were arguing over how money works and how to make more of it, we were quietly replacing human labor with fossil fuels in every industry we could imagine. And nobody noticed. We didn't update the model. And

---

† This is pretty much the foundation of what we still use today.

fossil fuel labor, as we know, costs almost nothing compared to paying an actual human being. And in our labor theory of wealth: **labor is the most valuable input.**

When the factory replaced a seamstress with a machine, labor was still an input into the model; it was just infinitely cheaper. When human labor gets replaced by oil? POOF! It's gone. When human labor gets replaced by a robot? POOF! It's gone. What did it turn into? PROFITS! We just invented some more money. When we plug this dirt-cheap energy into our circular-flow model, it's a rounding error compared to paying an actual human.

But rather than understand what was happening, or the real costs associated with using and burning all that energy, we decided instead to tell a story that our rapid development was the product of sheer human ingenuity. And I get it. The whole time this was all unfolding around the late 19th and early 20th centuries, shit was so flipping exciting. So many things were being invented. Can you even imagine getting your house electrified for the first time? Seeing a motherfucking car? Not having to make candles for two straight weeks every year???? It was the birth of the Incredible Everything Machine. And the narrative that humans were magnificent started to morph into feeling like we were invincible. So we just kept saying how "human productivity is increasing" without accounting for the very obvious fact that human productivity had been replaced by technology that was sucking up more and more oil and materials and dumping out more and more waste.

But the system works! Everything is the best it's ever been!

I knowwww! But just because everything IS the best it's ever been doesn't mean it can just continue to get better and better. This is a logical fallacy. You can't draw a straight line out from any chart and assume past performance guarantees future results.

There is a hilariously terrible book on the bestseller lists right now called *Superabundance*. In it, authors Marian Tupy and Gale Pooley argue that because, up until now, more humans have made more innovations that continue to make everything cheaper and better and more accessible to more people, that this absolutely will continue forever. And the more people we have on the planet, the better! Here's a quick summary of their findings from the intro to their book:

> Going back to 1850, the data collected by Tupy and Pooley show that for more than a century and a half, measured by time prices, resource abundance has been rising at a rate of 4 percent a year. That means that every 50 years, the real-world economy has grown some sevenfold. Between 1980 and 2020, while population grew 75 percent, time prices of the 50 key commodities that sustain life dropped 75 percent. That means that for every increment of population growth, global resources have grown by a factor of 8.[34]

The key here is that Tupy and Pooley aren't measuring abundance in availability of resources: they're measuring abundance in terms of how much things cost.

I know everything is wonderful and cheap and life is easy (unless you're one of the very many people for whom it is not), and so many people have been pulled out of poverty, and we have all the DiGiornos! But why do you think everything has gotten almost infinitely cheaper over the last fifty years?

The magic of capitalism, of course! Globalization and increased specialization have made the most efficient market machine!

But why *exactly* has globalization made things so much cheaper? Is it just that the spinning jenny means I can make eight scarves in the time it used to take to make one? Orrr is it that globalization means you can pay a child in Malaysia ten cents an hour to make your scarf or your sneakers instead of a fully-grown human subject to those pesky Western minimum wage requirements? And that very cheap human child is operating a very efficient machine that runs on—say it with me now—super cheap, abundant oil. And if we can take the human out of the equation entirely? Even better! And the more machines we've made that replace more humans along the way, the cheaper those products can be. You know that Malaysian child is out of a job the second a robot learns how to make a Nike.

Instead of understanding *how* everything got so much cheaper, Tupy and Pooley decided to infer from their data that more people on the planet actually *create* more resources. They don't think people use resources; they think we're a source of them.

Because they are conflating abundance with affordability using time prices, or how many hours you have to work to buy that iPhone. Finding more lithium makes lithium cheaper, which makes the iPhone more affordable for everyone,[‡] but it doesn't put more rare earth metals in the ground to make more phones.

Because material resources are only represented in our models by money, it's probably important to understand how exactly we decide to price these things. Since lithium is the rare earth metal upon which rests our entire plan to save the world, let's dig into the dynamics there.

The entirety of planet Earth has about 88 million tons of lithium that we know about, but only about 25 percent, or 22 million tons of that, is economically viable to mine. Okay, so if demand is going to increase to 4 million tons per year for all our electric cars and batteries, then we have like five years left? Ten?

Probably even more! The great thing about lithium is that, since we didn't really care about it that much a few years ago, no one was that interested in finding more of it. There were mines, they made it, they shipped it around the globe, and we were good. But now that we're talking about making every car an electric car, and every electric vehicle battery requires somewhere between 16 and 138 pounds of lithium, we just keep finding more and more of the stuff every year. Tupy and Pooley were right! In 2008 we thought there were only 13 million tons on the planet, and now we've got almost 90! And oh shit, India just discovered another 5.9-million-ton deposit! What's that? They just discovered

---

‡   Or, much more likely, the phone stays the same price, but Apple's shareholders get richer.

another massive deposit even bigger than that one! At least now we don't have to rely on the salt flats of Argentina, Chile, and Bolivia for all our lithium needs.

But there's a weird math that happens here. Since we just found so much lithium, the price dropped dramatically. Which means the amount of lithium on the planet that's economically viable to extract . . . has gone down. But when the easy resources are used up, and demand is still high, then eventually the price will go back up and those harder-to-access deposits will become economically viable again! What a weird system we've created.

But no matter how expensive it gets, and how many additional deposits become "economically viable," the amount of lithium available still isn't infinite.

And while we *can* recycle a lot of these things, we're just . . . not very good at doing that when everyone thinks forever is a real thing and not like, fifteen years from now. And recycling, because we live in capitalism, is also based on market forces. People will only recycle the lithium if the market price makes it profitable.

Even though there is less than one gram of lithium in a cell phone, finding the batteries to be recycled became a profitable endeavor when the price of up lithium increased 300 percent between 2016 and 2018.[35] But wait, oh shit, now the price completely collapsed again; it's down 72 percent so far this year! So lithium recycling isn't profitable? And we're just gonna keep throwing it away? Unless the price goes back up, and then we'll go dumpster diving for more old cell phones? Even though it's definitely finite and our entire plan for saving the planet from global warming hinges on making things that run on batteries that we can't make without lithium? Okay, just checking.

Much like with the oil, *when* the lithium runs out isn't the question. The question is: how expensive is it going to get when it starts getting scarce? And of course, it will still run out one day, because it's finite. But then we can just start pulling lithium out of the ocean because it's superabundant in there! Except we don't actually know how to do that yet. And it's probably still gonna be pretty expensive. But whatever, we're humans! We'll innovate! Materials science is always developing, and hopefully we can find another way to make batteries because the ones we've got now sure are problematic.

Despite everything we know about how our economy works and how our planet works and how much oil there is left and how much lithium there is left and how many electric cars we need to make, economists *still* don't bake that into the price of lithium or oil or coal. Even though we know we don't have enough lithium or neodymium to meet future demand. Even though we know the oil is going to run out.

We let the market we've created tell us how much things should cost, then we let the cost of things tell us all the decisions we should be making without questioning whether our models are based on the right assumptions. No one has ever thought that non-renewable things might be worth more than their easily renewable counterparts. And we don't include the cost of child mining or environmental destruction in anything at all. It's just how much it takes to make or get the thing.

Economists don't care if your scarf is made out of a renewable resource like sheep's wool, or an unrenewable one like polyester. The only difference between powering your scarf factory with oil or coal versus solar or waterpower in our models is their relative cost. How much they cost is based on how much they cost to get. Solar power costs what it costs to build the panels. Oil costs what it costs to suck it out of the ground (as well as how much demand there is, and how scarce we perceive it to be). Even though oil and coal and lithium are finite and destructive and sun and wind and wool are abundant AF. Their "value" is based on how much they cost right now to make, at this very moment. And the cheapest input is always going to be the preferred one, otherwise you're being "irrational." If it's cheaper to use oil than solar power, no business is going to switch. If it's cheaper to make a new hourglass than recycle the pieces, no one will recycle it. If it's cheaper to replace your dishwasher rather than repair it, people will keep throwing away entire dishwashers every time something breaks. No sane businessman would pay a human to pick his radishes when a machine can do it for a tenth of the price.

But the only reason any of these equations work is the dirt-cheap oil (and coal) that powers the machines and mines the lithium and ships the products from the countries with the depressed economies where they're being made for pennies.

We've drastically, catastrophically underpriced the most critical, irreplaceable input not just to our economic models but to our entire society. And for two hundred years, any voices of dissent or any calls to maaaaaybe include energy as a real input or cost (or raw materials even) have been silenced to serve the endless growth machine.

So, here we are today, where everyone—even the smartest minds in the field of economics—has accepted a circular model of labor and capital where goods and services exchange endlessly between companies and households, without exhibiting any physical relationship to the natural environment.

We're trapped in a system where all of economic policy is based on theoretical money built to encourage growth and maintain stability in systems we literally invented, but none of it is tethered to the natural world that we literally can't live without. And our main way to measure success is not only completely detached from human happiness or wellbeing—arguably the two things we should have been looking to maximize all along—it *requires* the destruction of the system we depend upon to survive. And our only plan to save the world is completely reliant on a bunch of incredibly rare, finite resources that we keep throwing in the trash. And we're looking at a very near-term future where the price of the irreplaceable substance that keeps the whole thing humming starts climbing real fucking fast.

Well, when you put it that way . . . yikes.

But people have been predicting this forever! And they've always been wrong!

You're absolutely right. And I hope I'm absolutely wrong. Someone, somewhere has always been shouting about the finite nature of the ball we live on and the impossibility of endless growth. And it doesn't help my case that everyone who has predicted the upper limits of human innovation and food production and population . . . has been laughably wrong.

Thomas Malthus was super worried about overpopulation decimating resources (even back in 1798) with the population boom that
accompanied the Industrial Revolution. At the time, the population of
Britain was around 7 million people. He predicted that food supplies
could continue to grow to feed up to 21 million people, but after that,
food production couldn't keep up with exponential population growth,
and a whole bunch of people would die sad, hungry, miserable deaths.
For the record, the current population of Britain is 67 million people,
and they don't seem hungry at all.

But nobody save for a few crazies like Malthus ever thought we
would actually hit a point of running out of stuff. We didn't have to
worry about the blue square getting too big for the green circle because
we weren't big enough to stop the system from regenerating. Because it
will regenerate—as long as you don't use too much.§ Our pre-industrial
or even steam-powered economy with a global population of around
one billion was just fine.

As that square grew bigger and bigger, anybody who suggested that
we might grow too big for our britches and beyond the capacity of the
system we rely on was dismissed as a doomsayer. I mean, everybody
is getting richer, just shut up already. And besides, plenty of people
believed one day we'd naturally stop growing once everybody got rich
enough. Adam Smith and equally important classical economist John
Stuart Mill didn't even see infinite growth as a real possibility or the
natural trajectory of the market. They both believed that all developed
economies would reach 'secular stagnation,' where there is nothing
wrong, but we just don't need to grow anymore, so we don't. And that
was a good thing. Everyone's needs would be met, and we'd all spend our
days writing poetry and lying in hammocks. Lolz.

Of course, the world we're looking at and living in today is absolutely nothing like the one where all these theories were invented. And
because we innovated our way out of every other limit, it makes sense
that we would just do that again forever and ever.

In the 1968 book, *The Population Bomb*, Paul and Ann Ehrlich

---

§   And as long as the stuff you're using is renewable, obviously.

famously declared, "The battle to feed all humanity is over," and predicted that hundreds of millions of people would starve to death in the '70s and '80s. And once again, they were right in theory and so very wrong in the numbers. Human innovation strikes again.

I don't want to discount all of human ingenuity. We have discovered things and created things that increased abundance on the planet. They've made our work easier without relying on fossil fuels to do it (occasionally). And that's great. You can apply those time prices to things like crop productivity and say yes, we're making more crops with less work on the same amount of land. We are better at getting our crops to produce way more, so wheat can be way cheaper, and even thought of as more abundant.¶ But you can't apply the same logic to un-renewable things. We can grow more wheat and wool every year to a point, but we can't grow more oil. We can find more lithium or cobalt in the ground, which makes it cheaper because of the way our economic system works, but it will never put more lithium and cobalt in the ground.

This innovation and abundance argument is the same one you'll hear every economist and technocrat spouting in support of infinite growth. All those doomsayers were wrong because they just couldn't imagine how much more humanity was going to innovate! And we have to keep growing because getting people out of poverty allows them to innovate even more! The fewer hours you're working to survive, the more hours you can spend working to figure out how to increase wheat yields even further! It's a virtuous cycle! Humans create resources! And the whole finite planet thing isn't an issue because of the First Law of Thermodynamics: energy is never created or destroyed. Screw entropy! Somehow we're just gonna turn all those landfills and carbon dioxide clouds back into food. Things may seem pretty bleak at the moment, but we're gonna innovate our way right out of this one too!

And I'm sure they're right—to a point. I can't wait to see all the exciting innovations that come out of the next three decades as we're faced with an extra billion people to feed on a severely warming and destabilized planet. It's gonna be so fun! Actually, you can just put this

---

¶   I'll be nice and disregard the whole industrial farming/natural gas fertilizer thing for a moment.

book down now. Why bother taking any actions at all to avoid potential global collapse when surely someone is gonna innovate us right out of it?!

Somewhere along the way, in the mix of all these white dudes trying to figure out how money works and laughing at Thomas Malthus, we sort of . . . forgot about those pesky limits of our ecosystem. We forgot that Malthus and Ehrlich were way off in their calculations but not in the core underlying assumption: we can't make an infinite amount of food or stuff.

There is no way around it. Our economy takes materials. Lots of materials. And we need energy to get those materials. And if a thing doesn't take materials, it takes energy. We need energy to do literally everything. And the renewable energy that's supposed to innovate us out of running out of oil takes even more materials. A lot of rare materials that are very, very hard to recycle. And we're still only basing the cost of those materials on how much it takes to extract them, not how much value they have to our society.

And the bigger we grow our economies, the more materials and energy we need, the bigger the blue square gets. Almost all of the innovations that have saved us time and time again are based on using more energy. Energy that we don't have to spare. And even if we figure out nuclear fusion or deep geothermal at scale or finally make a battery that can power a city for a whole day, we still need all the raw materials to build all those fusion plants and giant batteries. And no matter how good we get at recycling materials, they are always going to degrade and always need to be replaced because of the law of entropy.

My best hope for salvation is stumbling across a giant asteroid that happens to be filled with every rare earth metal we need, and then *Armageddon* style, we send Bruce Willis and Ben Affleck up there to go drill it all out. What? It could happen.

Too Big for Our Britches

Under the discipline of a steady-state economy
free energy would be a blessing; within the context
of a growth economy it would be a curse.

—Herman Daly

HOLY SHIT OH MY GOD WE FOUND THE ASTEROID! AND WE PERFECTED DEEP GEOTHERMAL! And we're using it to power our clean hydrogen fuels that don't pollute at all. And this incredible hydrogen fuel powers all of our airplanes and freight trains and Mack trucks and heavy machinery and everything else. All of our problems are solved! Let's all start taking private jets to work!

Just kidding. The problem with finding even more energy and more materials is it's actually a bad thing. It would be great if we could replace fossil fuel energy with a totally clean and carbon-free alternative. We still realllly need to do that, obvi. But finding an endless source of energy and an asteroid full of materials so we can keep the world population growing and developing would kill the planet. End of story.

There are plenty of wild arguments out there saying "the planet can sustain billions more people." But growing the food and keeping the ecosystem in balance are two totally different things. If we found a near-boundless source of energy like deep geothermal, all we would do is keep growing. Our planet is a closed system that requires balance. It always has been. And endless human population growth or economic growth—or endless growth of any kind for any species—simply isn't an option.

When we built all the incredible machines that run on the incredible oil, it allowed our global population to explode. It took us almost 300,000 years as a species to reach one billion humans in the early 1800s . . . and it took us just 123 years to double it to two billion in 1927. The next doubling to four billion took just 47 years. And we doubled it *again* in 2022. In fewer than 200 years we multiplied by EIGHT. Since I was in high school, we've added another two billion people. No wonder everywhere feels so crowded. If you understand exponential growth, you can see why this is a problem (and what Malthus was so concerned about).

Somewhere while we were creating economic models built on the premise that natural resources are free and that our ecosystem is infinite, we came to the logical conclusion that humanity could just keep reproducing and chopping down forests and burning oil forever and absolutely nothing sounds weird or wrong about that. Economists will look you dead in the eye and tell you the planet can hold eighty billion people.

You can argue all you want that technology will solve all of our problems, and I'm sure the technocrats are right to some degree. I'm not a Luddite, and I'm sure we'll get more efficient in a lot of ways. It just can't go on forever. It never could.

You know what infinite means, right? Absolutely no limit. There are eight billion of us on the planet now, but if you had to guess, how many people do you think it could hold and feed? Surely the number isn't "infinity?" Is the entire planet going to become four-hundred-story skyscrapers? Even if Malthus and the Ehrlichs were wrong, surely there is some population of resource- and energy-gobbling humans that would be the maximum, no?

Scientists obviously disagree on how many people the planet can sustainably support, but it's really a formula that depends on how richly people are living. The more resources a human sucks up, the fewer of that type of human you can have. Can the planet hold ten billion people living relatively simple lives? Sure! I think so, anyway. Can it hold ten billion people living like Americans? Definitely not. According to plenty of ecologists, we are *already* in a state of overshoot. It's not going to happen, it's happen*ing*.

Overshoot is a very common phenomenon in nature. A species, say a fox, gets really good at hunting rabbits. So their population grows as they eat every rabbit they can find. They multiply like, well, rabbits. But then there aren't enough rabbits left to feed all the foxes anymore. So what happens? The foxes start to die out because they're starving to death, and the rabbit populations rebound when there aren't a million foxes eating them, and everything is right with the world. Right now, we're the foxes.

The UN predicts that global population will reach 9.4 billion by 2050 and will max out around 10.4 billion in the 2080s[36] (as industrialized economies always see vastly decreased birth rates). So adding another 2.4 billion people to the planet would be a 30 percent increase, nbd. There is plenty of room for 30 percent more people. Just look at all that land in Africa! And the middle of America and Australia? Hardly anyone lives there! Ooh! And Canada and Greenland! They're bursting with uninhabited land!

True. But of course, physically putting the billions of people onto the planet like trying to shove a week's worth of clothes into your "personal item" on a Spirit Airlines flight isn't really the issue at hand. It's the feeding them . . . and clothing them . . . and giving a bunch of them cars. And the more comfortable, developed lives they lead, the more land and resources and energy they gobble up. A woman in Africa having five children while living a subsistence life is a whole lot more sustainable than a bougie, childless couple in San Francisco. Turns out, every person takes up a lot more space than their house.

Does my footprint look fat in this apartment?

Speaking of which, if you were to guess, how much land do you think your life uses? To farm all the food you eat, and grow the cotton that makes all the clothes you wear, and to refine the oil into plastics to make the 150 pieces of packaging you throw away every week? What about the factory where the new couch was made that you bought to "freshen up" your living room? As we just discussed, all that land—all that stuff—has to come from somewhere.

One way we can look at this is global hectares. A global hectare is 10,000 square meters, or about 2.5 acres. For reference, the average plot of land for a home in the US is about a quarter of an acre. Each hectare then, is the space of around ten suburban American homes and their perfectly manicured lawns. It's the street you grew up on.

Globally, the total number of biologically productive hectares of land and water is around 12.2 billion. With eight billion people, that's about 1.5 hectares per person if we all split it equally. Seems like an entire neighborhood of land is enough to support one person, right? No? Two neighborhoods? Five? Oh shit, how many neighborhoods is it taking?

Surprise! Each American uses about 8.1 hectares of land to live their life, while the country itself has only 3.4 available per person.** Here in Spain, the average is much lower. We only use about 4.4 per person—but in a country that only has 1.5 available. And of course, in many countries in Africa where basically no one has a car and they still live largely on subsistence farming, they're well within planetary boundaries at a single hectare per person.

And cities? Cities are fucking nuts. Every city in the world uses 100–1000× more land to support the people in the city than the land it sits on.

Obviously the biggest (and most important) land use is food. The story of everything is the story of energy, and as of today, we've converted 38 percent of the global land surface to farming and cattle grazing. When you drive around America, how much land is already being used for farming? Like, all of it, amirite? Same is true here in Spain. You can't find even a corner of a hill that doesn't have an olive farm on it. Just looking around, I'm not sure where we're putting all these extra farms to keep our growing population fat and happy.

There's a term in shipbuilding known as the Plimsoll line. And that

---

** You can use this fun calculator to check out your own ecological and carbon footprints: https://www.footprintcalculator.org/.

line is how much stuff you can put in the boat before the boat sinks. If the Plimsoll line drops below the water line, you're fucked. So when you're packing your ship up with all the cargo, you just make sure that Plimsoll line stays above the water. Well, Earth has one of those too. Along with every system in existence. It's called carrying capacity.

And according to plenty of scientists (and zero economists), we've flown right past it. A 2002 paper published in the journal *Proceedings of the National Academy of Sciences* estimated that humanity's load corresponded to 70 percent of the capacity of the global biosphere in 1961 and grew to 120 percent in 1999.[37] I don't even wanna think about what it's at now.

Do my cows look fat in this hectare?

Let's go back to thinking about the Earth as a single, closed ecosystem. Actually, let's never stop thinking about it that way, because that's what it is.

I want you to imagine all the plants and animals in every stunning mountain range and vast desert and the depths of the oceans. Imagine every creature crawling, flying, sleeping, eating, and majestically existing. Imagine all the trees and grasslands they exist in. Imagine all the humans and all the cows and chickens and pigs we eat. Imagine all the farms growing all the crops that allow us to keep living. Now imagine how much they all . . . weigh.

Every living, breathing thing you just imagined is part of what ecologists call biomass. It's the measure of the weight of all living things on Earth. Humans, for example, make up just 0.01 percent of all life on the planet. Plants are 82 percent. If we just look at the biomass of mammals, humans account for 34 percent.[38] The livestock we raise to eat makes up 62 percent. That puts us and our cows and pigs at 96 percent of all mammals on the planet, leaving a paltry 4 percent for wild animals.

And birds? The birds we raise to eat are 71 percent of all the birds in the world, more than twice the biomass of birds in the wild.

> When I do a puzzle with my daughters, there is usually an elephant next to a giraffe next to a rhino. But if I was trying to give them a more realistic sense of the world, it would be a cow next to a cow next to a cow and then a chicken.[39]
>
> —Professor Ron Milo

If those numbers don't sound weird or terrible to you, just keep in mind that from 1900 to 2015, the hundredth of a percent of mass that is humans drove an 85 percent decline in the biomass of wild mammals on the planet. Seems like an outsized effect, no? And we started this long before oil. Oil just sped it up by a few millennia. From 7000 BCE onward, as farming and herding economies spread around the globe, we converted millions of acres of forests to grasslands and grasslands to deserts. Ecologist R. O. Whyte estimates that the forests of Anatolia (modern-day Turkey) were reduced from 70 percent to 13 percent of the total surface area between 5000 BCE and the recent past.

Surely we need to keep some forests . . . somewhere? I can't remember why exactly, but I feel like trees are important for something. Oh right! Oxygen! Trees and plants physically take carbon dioxide out of the air, use the carbon to become bigger plants, and give us that oxygen back we love to breathe. What happens when we clearcut forests and lose the carbon sink†† of millions or billions more trees? What happens when we take a wild land or a forest or a meadow or a mountainside and turn it into a massive monoculture wheat farm? And what happens to the habitats for all the animals that live in them?

When we clear the land for farming, we destroy the biodiversity. No more bugs and birds and whatever else was living in that forest before. What do you think they're doing in the Amazon right now?

But even before farming, let's not forget our cave-dwelling ancestors were the ones who killed off all those mammoths.

---

†† A carbon sink is something that sucks carbon out of the atmosphere and holds it, as opposed to releasing it, which is a carbon source. Forests are carbon sinks. Seagrass in the ocean is a carbon sink. Soil is a carbon sink. We love carbon sinks.

Changing distribution of the world's land mammals

Mammals are compared in terms of biomass, measured in tonnes of carbon

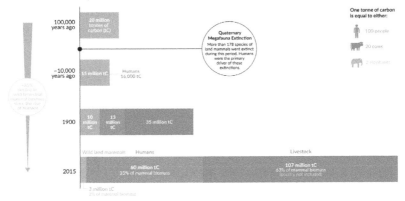

Note: Estimates of long-term biomass come with significant uncertainty, especially for wild mammals 100,000 and 10,000 years ago.

Data Source: Our World in Data | Licensed under CC by the authors Hannah Ritchie and Klara Auerbach

The wildest part about this graphic isn't how much the wild biomass has decreased, it's how much our human and domesticated animal biomass has *in*creased: from millions to billions. We've multiplied nearly ten-fold. Since nothing can be created or destroyed, where is all that biomass coming from???

Well, when you have a baby, you're not pulling it out of thin air . . . you're "eating for two." You're making that human out of food—out of energy. I know, it sounds weird. But it's facts. As Marvin Harris loves to remind people, we will use every excess calorie we can get our hands on to make as many babies as we can. Because, you know, we're basically just bugs. But how did we magically invent so much food energy over the past 150 years?

Does my nitrogen look fat in this soil?

So far, we've gotten all those extra calories by continually converting wild land into farmland (destroying all that beautiful, wild biomass and decimating those delicate ecosystems) and continuing to increase our crop yields over time. And boy are we good at increasing crop yields. No wonder no one's worried about feeding all these future humans.

In the US, we've seen cereal yields increase from 2.52 tons per hectare to 8.27 tons since 1961.[40] That's a 328 percent increase! Looking at the world as a whole, we've gone from 1.35 tons per hectare to 4.15, a respectable 307 percent. Based on this, in another sixty years we could triple it again. Surely, we can increase crop yields ad infinitum! Everybody get to baby making!

Except, there's one small caveat. How we got it done. Back in the day, increased yields were due to figuring out better farming techniques. Like how to plant things at the right time and get more efficient at harvesting them, you know, basic stuff. Then we started breeding crops that were more productive. Awesome. The same plant can make 20 percent more food than it could before using even less energy. Cool. Then we engineered crops that were resistant to pests and viruses, so we lost fewer crops. But once you master these basic things, there's really only one way we've figured out how to increase yields: pumping our largely degraded soil full of nitrogen-based fertilizers in order to keep producing. The only reason farming even still works in a lot of places is because of the seemingly endless supply of fertilizer from fossil fuels that we pour right back into farmland that would otherwise be completely dead. The second we stop fertilizing, it's gone. The farmers know this, that's why they have to keep buying the fertilizer!

Remember that wild chart from earlier? Fully half of all global food production would be impossible without fossil fuel fertilizer.

Here is another fun chart showing cereal yields compared to fertilizer use. Surprise! Increased crop yields are pretty much directly correlated to fertilizer use.

Cereal yield vs. fertilizer use, 2020

Yields are measured in tonnes per hectare. Fertilizer use is measured in kilograms of nitrogenous fertilizer applied per hectare of cropland

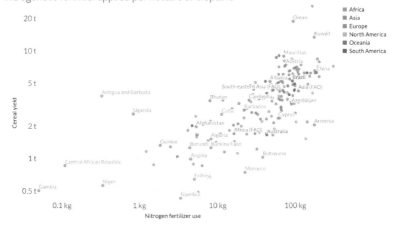

Data Source: Food and Agriculture Organization of the United Nations   OurWorldinData.org/fertilizer | CC BY

Okay great, so we keep pouring more fertilizer on, and we'll have corn stalks twenty feet tall!

I hate to state the obvious, but even with unlimited fossil fuels, there are biophysical limits to crop yields. One corn plant could maybe yield three ears of corn instead of one, but it can't provide twenty-seven. As developing economies get better at farming (and use a whole bunch more fertilizer) they see their crop yields increase on a mostly linear basis until . . . they don't. It plateaus.

According to a paper in the renowned journal *Nature*, both rice in eastern Asia and wheat in northwest Europe, which account for a third of global rice, wheat, and maize production, have recently hit yield plateaus.[41] Since 2002, food production growth has been increasing mainly by expanding habitat, not by further increasing yields. Crop production area has increased by nearly ten million hectares per year. Unsurprisingly, nearly all of the increased crop area since 2002 has occurred in rapidly developing countries in South America, Asia, and Africa.

So . . . if we're gonna increase the population by about 30 percent—and the 85 percent of the world still living in developing economies wants to start eating more meat because meat is delicious and now they have enough money to buy it—where are we putting all those animals? There are a billion Chinese people coming out of poverty who really love eating pork. The demand for meat livestock is projected to more than double by 2050, far outpacing population growth.

How much more land do you think we can convert? If we're already at 38 percent of land globally, what's the cutoff? Fifty percent? Sixty? Or maybe we already passed it. No matter which way you slice it, we can't keep this up forever. We can't cut down 80 percent of the wild ecosystems on the planet and have the planet still regenerate enough to support us.

And while we already make enough food to technically feed our projected max population of 11 billion people, only 55 percent of what we grow gets eaten by humans. The rest goes to raising those precious cows and pigs and chickens.

Now, are we at the cusp of creating lab-grown meat? We are! And I'm totally on board with this. I'm sure there are a lot of other concerns like, is it even healthy, or maybe living in balance with nature is a better idea than continuing to mechanize everything. But whatever, I'm here for it. It takes 1,800 gallons of water to produce one pound of beef, which is absolutely insane when you think about it. Eighty percent of all that land we've cleared is reserved for these majestic beasts. If all meat was grown in a lab, it would reduce that land use by 99 percent![42]

But, since things don't get created out of nothing, how do you think you get that meat to grow? Just put it out in your garden like a rosemary bush? Of course not! We use fossil fuels! Duh!

Or maybe you're not worried about any of this. Maybe you really don't believe that humans depend on the environment. Maybe you truly believe that we can manipulate crops and grow cows and totally decimate forests and grasslands for the rest of time. If you really really think that planet Earth will continue to produce ever greater amounts of food no matter what we do, then I guess you really can put this book down. Because that's totally fucking nuts.

Does my carbon look fat in this atmosphere?

OH MY GOD! YOU WERE RIGHT! We magically increased crop yields by 300 percent again, and we're all eating lab meat, and we built some very nice apartment buildings up in Greenland for those two billion people on their way. And the weather just keeps getting nicer up there. I can't believe I ever doubted you. We are the masters of our domain!

Oh wait, there's just one more thing.

Trash.

We make a LOT of trash.

Did you see when the garbage men went on strike in France for just a couple of weeks? That was a lot of trash. Now think about the trash for a whole year. Now think about the fact that Paris is a city of just two million people—two hundredths of a percent of the planet.

According to the World Bank, the world generates about two billion metric tons of municipal solid waste annually.[43] High income countries generate the lion's share, but the more developing countries join the ranks of Uber Eats and frozen pizzas, the higher that number gets. That 2 billion tons is projected to become 3.4 billion tons by 2050, more than double the population growth over the same period.*

But okay, two billion, four billion, whatever, we've got landfills. We can dig a bigger hole to put it in, or maybe compress it into a trash block? Or maybe we can burn it and generate energy while the CO_2 goes directly into some carbon capture thing? Is anyone working on that idea?

Honestly, I'm not that worried about the trash we throw in a can that gets picked up every week, and then we completely forget about it until there's a massive strike in Paris reminding us how bad it smells.

Why? Because the trash is a drop in the bucket compared to our favorite thing to throw away: carbon.

Remember how nothing is created or destroyed? Well, it's as true for plastic bottles and banana peels as it is for fossil fuels. When we burn those carbon-based life forms that were sequestered and compressed

---

* Notice how many things double far before actual population doubles when people start living more industrialized lifestyles?

in the ground a hundred million years ago, they release their pent-up sunlight energy and become carbon dioxide. They were carbon-based when they went in, and they are carbon-based going out.

And all that carbon . . . doesn't go anywhere. Except right back into our beautiful green circle.

The biggest source of human pollution by *weight* is carbon dioxide. That sounds weird, so think about it for a second. The human species spews out 37.12 billion METRIC TONS of carbon dioxide every year. I know, it's a gas, so it feels like a thing that's just floating around. But it weighs more than 37 billion tons. All that trash we throw away was just 2 billion. Total global plastic production annually is a paltry 460 million.

Annual waste in millions of metric tons, 2021

Comparing the weight in tons of plastic waste, municipal solid waste, and carbon dioxide emissions globally

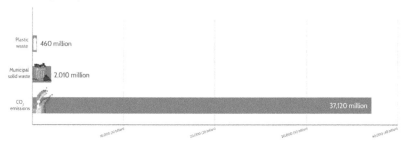

Data Source: World Bank, "Trends in Solid Waste Management," and the Global Carbon Project (2023)

Our CO_2 emissions weigh 80× more than all the plastic produced annually in the entire world and 18× more than ALL THE TRASH WE THROW AWAY.

And the making of the food we eat is one of the biggest contributors. Agricultural production, including indirect emissions associated with land-use change like cutting down trees, as well as methane from cow farts accounts for 25 to 35 percent of total human-generated greenhouse gas emissions.[44] Farming releases more carbon dioxide than all our cars, trucks, trains, and airplanes combined! But suuuure, let's create another two billion people and give 'em all DiGiornos.

According to ecological economist William Rees, climate change isn't even the problem. The massive amount of pollution—both plastic bottles and dark smokestacks of carbon and car exhaust we're shitting into our own beds—isn't exactly the problem. Climate change is just the downstream effect of growing too big for the system we exist within. The system is renewable. It can grow more food. It can turn CO_2 into bigger trees that turn it into oxygen. Every system that exists on the planet exists because it was created that way by hundreds of millions of years of co-evolution. The planet isn't a milkshake that once we suck it down to the bottom there's no more left. It's a milkshake that's slowly being refilled, but we've got a hundred different straws in it and we're drinking it way too fast.

We are in a state of overshoot because we're using more resources than the land can renew. We're eating the fish faster, we're cutting down the trees faster, we're degrading the soil faster, we're killing the land and habitats by raising too much cattle faster—and we're obviously using the oil faster—than our closed ecosystem can regenerate. Finding an infinite energy source, even one that doesn't pollute, would just mean even more people sucking down that milkshake even faster. And we can't fertilize our crops with nuclear fusion.[*]

There's a helpful way to think about this that has to do with basic finance. If you have a million bucks in the bank, it might earn around $40,000 in interest a year. You can live on that. You never touch the million dollars, and every year it provides another $40,000. Amazing. The Earth is the same way. Every year crops will grow, and fish will have fish babies, and cows will have cow babies, and trees make fruit babies and tree babies, and everybody's happy. It's circular flow, and it's regenerative. But you run into a problem when you start spending more than the interest that's accruing. If you spend $100,000 this year instead of $40,000, your principal balance goes down by $60,000. And the next year, you don't earn as much interest because it's based on how big your principal balance is. And then every year you spend more than the interest, it gets lower and lower and harder to live on the interest, until

---

[*]   Or maybe we can? But it still won't solve our problems.

one day you're broke, and you gotta ask your dad to give you another trust fund.

When we think about global wealth from this perspective, it's not an accumulation of stuff, but a capacity to reproduce next year what we already consumed this year. This is the Earth. And right now, we're on a spending spree. She can't regenerate fast enough every year. So the principal balance is getting lower, and there isn't another trust fund to tap. There is one thing we definitely can't make more of, and that's land.

We can't even hold the people we already have.

And we're still fucking growing. Constantly. And if we ever stop growing, the whole system we've created collapses.

Any system driven by positive feedback self-destructs. Positive feedback loops allow unprecedented growth—more oil means more food means we can have more babies who need more food who make more babies—and on and on. But when those positive feedback loops lead to overshoot, the system will correct itself.

Like when our ancestors killed out all the mammoths and had to start killing their babies and going to war, and when the Aztecs over-farmed the soil and it stopped being productive, overshoot corrects itself. Human overshoot has killed populations of humans throughout our entire history. It's a big part of what kept us in check as a species for 300,000 years. The self-destruction is a feature, not a bug. But since we discovered oil, we found a way around it, and we've been singing infinite growth ever since.

We act like we can just keep pumping the ground full of fertilizer and clearing the forests and everything will be okay. But even if we invent a fertilizer that isn't made out of natural gas, the Earth doesn't just churn out bell peppers and romaine lettuce for your salads because you asked nicely. Nutrient-rich topsoil is essential for growing the food we eat. And when you over-plow, overgraze, and overplant, that soil degrades. It degrades when you take too much and don't let it regenerate. And then . . . it doesn't grow things anymore. Ecosystems are delicate and complex. Soil degradation is often irreversible. And no amount of fertilizer can fix it when it's gone.

You can argue we'll innovate a million different things to avoid disaster, but we still rely on the ground to grow plants that use the sun to become food. Food for animals and food for us. And if the ground can't make the food we eat, if the ecosystem can't support the plants and animals that make life on Earth possible—we die.

Our Plimsoll line is below the water line; the ship is sinking. It's not "getting there" or "going to happen." We're already fucking there. The whole time we've been enjoying this superabundance and unprecedented global peace for a full seventy years, we've been packing the boat with more and more cargo. I mean, it's been great. It's been a wild ride. We've lifted billions of people out of poverty. Hell, we've *created* billions of people, and most of them didn't die of starvation. That's pretty cool. We don't have to poop outside or use that poop to fertilize our fields (though we probably should be) or spend two weeks a year making candles. I'm all for the ease and convenience of modern life. I wish it could go on forever. It just can't. At least not in the form we've come to know as the only way things are.

We've enjoyed our superabundance of energy as much as humanly possible, but it could never be infinite. And whether it's ten years or fifty years or a hundred years, it doesn't really matter. Because the party has to end. For the record, I think we're going to start seeing the cracks pretty soon here. Hence why I'm writing this book.

Think of it like a hangover. For the last hundred years, we've been having one helluva rager. Partying like it's 1999. Partying harder and harder with each hour into the night. Tequila shots, molly, cocaine off a sex worker's tit, you name it. But now the sun is about to come up, we've done all the drugs, the house is trashed, and we're gonna have to pay the piper for all those good times. And if you've ever had a hangover after a massive bender, you know it is not going to be pretty. And the bigger the party, the longer it takes to get the house cleaned up.

This period of time will go down in history. It will also be ... a period of history. There is no world where this world is still this world in a hundred years. Or even fifty years. And it's probably a whole lot less than that. So we can either pretend like that's not happening and let our children and grandchildren deal with it. Or we can do something. We can start telling people that the party's over, and it's time to go home. Instead of cracking one last beer at six a.m., we can start cleaning up the mess we've made.

.

How bad could it be, really?

Pretty F*cking Bad

I want you to act as if our house is on fire. Because it is.

—Greta Thunberg

Okay, so let's say we ignore the whole "carrying capacity, planetary boundaries" thing and just keep on trucking because we believe in the human power to innovate our way out of anything. Because let's be honest, that's probably what's going to happen anyway. But if we keep on going the way we've been going, there are two different strains of problems we're imminently facing.

One is the pressure put on society when an entire economy based on cheap oil shipping things all over the world suddenly has its main, dirt-cheap input messed with.

The other is a piece I ironically haven't really touched on yet: climate change.[†] Which, of course, is an externality directly caused by burning all that cheap, cheap oil. It's called an externality because the cost of climate change is not currently included in the price of oil, which it most certainly should be.

Both of these two strains will have and are having massive

---

† Read Peter Zeihan's book if you want to understand more about the collapse of globalization without all that pesky nonsense about global warming.

downstream ripple effects throughout our lives and across the entire global population. Resource scarcity in an infinitely expanding economy in a finite ecosystem is a very real problem, with or without the damage we're doing. But our fossil-fuel-powered economy *is* doing damage. And a lot of it.

As we just discussed, CO_2 may be a gas, but it still has substance. It still has weight. Most of the air we breathe is made up of nitrogen and oxygen, which do not trap heat. But the more of that CO_2 lives in the atmosphere, the thicker the air gets. Normally, when sunlight hits our planet, a lot of it bounces off and goes back out into space. This bouncing is known as albedo. Right now, we love albedo. But when the air is full of particles like CO_2 and methane (aka natural gas, cow farts, or CH_4), less of the sunlight escapes. We measure this concentration of carbon dioxide in the atmosphere in parts per million, or ppm. Currently, we're at 421 ppm of CO_2, which is 51 percent above pre-industrial levels (and wildly higher than any "natural" carbon cycles the Earth experiences). Here's an utterly absurd chart to help drive that point home:

The Keeling Curve
Global atmospheric CO_2 concentrations for 800,000 years

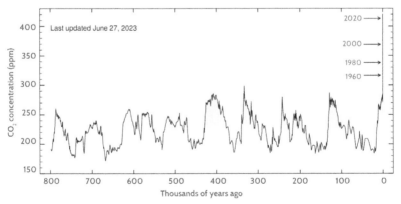

Data Source: Chart adapted from Scripps Institution of Oceanography, UC San Diego. CO_2 data starting in 1958 are from the Scripps CO_2 program at Mauna Loa, supported by the US Department of Energy (DOE). Ice core data prior to 1958 is from the National Oceanic and Atmospheric Administration (NOAA)

Yup. Looks totally normal to me.

The higher the ppm, the more of the warm, solar radiation can't bounce out and gets absorbed by our land and oceans. This is called the greenhouse effect, and it's why we call them greenhouse gases. You know, because greenhouses are built to trap heat. Different gases are better or worse at trapping heat than others. Methane, for instance, traps about 100 times more heat than CO_2, but it also dissipates from the atmosphere much more quickly.[45] Over 100 years, methane is about 25 times as warming as CO_2. But considering it does most of that damage in the first ten years, we probably should care a little more about the short term here. Hydrogen itself isn't a greenhouse gas, but it does some funky stuff in the upper troposphere that makes it harder for methane to escape, increasing its warming effect by multiples. So we reallllly don't want to release that either. So much for that green hydrogen fuel plan. And there are some other things that affect the amount of radiation that gets bounced off versus absorbed, such as the color of the land. Ice sheets are white, so they reflect more light. Asphalt is black, so it absorbs more.

And within these dynamics, we have a lot of positive feedback loops that create runaway warming. When the icecaps melt, there is less white space, which means less radiation is reflected and more is absorbed, which melts the rest of the ice even faster, and causes even more warming. Oh, and all that permafrost is currently a carbon sink. Arctic permafrost alone contains 1,700 billion metric tons of carbon between methane and CO_2. So as it melts, more greenhouse gases get released, which speed up warming even more, which melts even more permafrost, and we get even more methane.

The overwhelming consensus by scientists is that if we hit 1.5°C (2.7°F) of warming over pre-industrial times, we'll see catastrophic, likely irreversible shifts in our climate and ecosystem. Two degrees will be complete collapse. In case you're wondering, we're currently at 1.15°C, and with El Niño arriving this year, we're already touching 1.4°C, albeit temporarily.

I know everyone already knows this is a thing. We elder millennials have been learning about greenhouse gases since middle school. Despite the brilliant PR campaign by oil tycoons and big business to declare a very provable scientific fact a hoax, I'm pretty sure it stopped working when the climate started changing in our backyards. Spring comes in March instead of April, summers are longer and hotter and more miserable, heat waves are longer and more intense and more deadly, wildfires are unstoppable, polar vortexes are a thing now, and Texas fucking froze over. Texas. Of course, even considering all those things, nobody really seems to care. But since no one really seems to care about slowing emissions or burning less fuel or how many tenths of a degree the planet warms, I want to look at this from a slightly different perspective.

I want you to think about allllll of the oil we have in the ground. According to BP, right now, the entire world has 1.732 trillion barrels of oil in reserve. 'In reserve' means we both know about them and can access them.

As of 2021, we were consuming 94.1 million barrels per day, or 34.3 billion barrels per year. This lines up nicely with our "fifty years of oil left" prediction.

According to the EPA, each barrel of oil burned releases 426.1 kg of CO_2. Presumably this accounts for the part of the barrel being turned into plastics that trap the carbon for hundreds of years rather than becoming atmospheric carbon, but I'm not the one doing the math here.

So based on their math, if we were to burn every barrel we've got, we would release 738.2 billion metric tons of CO_2 into the atmosphere. Actually, let's call them gigatons because that sounds like an even bigger number even though it's exactly the same thing: 738.2 gigatons. And let me be clear—that does NOT include coal. That's just the oil. If you add all the coal we know about, that number jumps up to 2,795 gigatons.[46] Luckily for us, that last number was calculated in 2012, and I couldn't find a more recent one, so at least a big chunk of all that carbon is already busy doing its warming.

It's not easy to tell you exactly how much warming will happen from each barrel burned or ton released. It's far more complicated than any back-of-the-napkin math can show you. And this doesn't account for carbon emissions from deforestation or methane from cow farts or hydrogen leaks or anything else. And the carbon sinks of soil and forest and ocean continue to suck up around half of the carbon we pump out.

But if we already release 37.12 gigatons of carbon as pollution from fossil fuels each year, and we've already seen the planet warm by over a degree from the oil and coal we've burned in the past hundred years, and we've burned more than half of the oil we've ever burned in the last thirty years, and we only have 0.4 degrees left before shit starts getting real messed up, something tells me we're going to hit 2 degrees of warming before we have the chance to slurp around the bottom of that sweet crude milkshake. According to the Global Carbon Project, we only have ten years left at current emissions before we pass the 1.5-degree mark. If we started filling up a cup at the beginning of the Industrial Revolution with all the carbon we could emit before the cup overflows (and everything collapses) that cup would be 92 percent full. The Intergovernmental Panel on Climate Change (IPCC) estimates we can only emit another 420–580 gigatons (12 to 15 years of current global emissions) before the whole thing collapses.[47] And that was in 2019!

What's that? The World Meteorological Organization just predicted we've got a 66 percent chance of hitting 1.5 degrees in the next five years.[48]

It's like a fun game! Guess which happens first! Will oil become insanely expensive as reserves dwindle, collapsing our incredibly fragile economy? Or will we pass 1.5 degrees first, collapsing our incredibly fragile (though resilient) ecosystem? Will they happen at the SAME TIME?[‡] And what might this potentially dystopian nightmare look like?

Well . . . there are a wide range of possibilities, and where we end up depends largely on how much we're willing to do about it. But the thing is—a lot of this shit is *already* happening. And if we continue to burn every drop of fossil fuel we have just waiting for that green hydrogen plan to pan out, shit is going to get a lot worse. I mean, a lot.

---

‡   Check out the back of the book for a super fun "End of the World as We Know It" BINGO card!

The Short Term

I'm fucked. You're fucked. The whole department is fucked. It's the biggest cock-up ever. We're all completely fucked.

—Sir Richard Mottram, former permanent secretary at the Department for the Environment, 2002

When I say "short term" I'm talking about things that are most likely going to happen in the next few decades. Some of them are going to happen even sooner than that. And some of them have already started in one way or another, so the next five to thirty years are going to be a question of how much we see these things, which ones happen first, and how bad they get.

Extremely high gas prices

I want to take a brief moment to talk about something that has perplexed me since I was old enough to even vaguely understand the economy. Why, whenever gas prices change at the pump, do people blame (or credit) the American president for it? Gas is a global commodity. Events around the entire world dramatically affect the price of this substance the entire world needs to do ... literally everything every day. Sure, the American president can release strategic reserves to soften supply shortages and lower prices a bit, but he (or she) does not control the price at the pump. Also, if the President *could* make gas prices lower whenever they wanted, don't you think they *would*? Say it with me: The US President has no control over how much you pay for a gallon of gas.[§] When gas was recently over $5 a gallon in the US? It was over $8 a gallon in much of Europe. So stop blaming Biden or Trump or whoever else. It is a global commodity for which the price is set by global conditions of supply and demand. Here is a chart that shows you the direct correlation between the price of WTI crude oil and gas prices in the US:

WTI crude oil vs. US gasoline prices, 1986–2023

Comparing the monthly price performance of West Texas Intermediate (WTI) crude oil vs. national average gasoline prices back to 1986

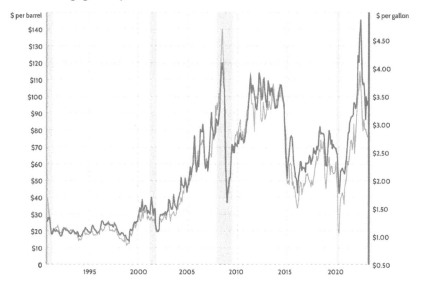

Data source: Macrotrends

---

§ Unless you want to nationalize the oil industry and use the profits to fund social welfare programs, which is a great idea! Let's do that! But the prices at the pump will still be contingent on global commodity prices unless we can source all of our oil domestically. They just won't *also* be contingent on price-gouging, profit-hungry oil companies.

Okay, now that we got that out of the way, let's get back to the whole running-out-of-oil thing. For the sake of argument, let's say we find a way to trap all the carbon and stop global warming in its tracks, and we can just keep burning all the oil we have left.

Thought $5 a gallon was bad? Try $25. $50? Or even worse, how about "no gas at all." If you think there is no way that will ever happen in your lifetime . . . you're probably right. We have reserves, and we will use them. If I know anything about how capitalism works, we will frack out every ounce of oil and natural gas we can find no matter how many earthquakes it causes or water supplies it poisons. But one day—with or without climate change—that will become economically unviable. If you're only twenty years old right now, that dry oil date starts to look a lot more plausible in your lifetime. But the "oil goes from $81 a barrel to $700 a barrel" scenario is wayyyyy more plausible for everyone who is alive and reading this.

But don't worry. We'll always still have electricity because if we ever run out of oil, we'll just go back to burning coal for everything. Which is what the rest of the world will do as well. In fact, it's already happening (I'm looking at you Australia and Germany). Which, of course, will severely exacerbate climate change as coal is the dirtiest thing you can burn (and the kind most countries have available is the dirtiest kind of all), which only destroys the environment further, so this could be a real fun one. As Peter Zeihan puts it: "We are completely capable as a species of devolving into a fractured, dark, poor, hungry world while *still* increasing greenhouse gas emissions."

But either way, as oil gets harder and harder to come by, as reserves take more and more money to get the oil out—money lots of countries won't have—the price of it is going to get higher and higher. This is not a theory; it is a fact. No one is denying that this is going to happen. Okay, some people are, but they are delusional. There are no more massive, easily accessible oil deposits to discover under the earth. We've found it all. It's just a question of getting it out, and who has the money and the know-how to do it. When the United States figured out fracking, we released billions of barrels of oil that previously couldn't be extracted because it existed in tiny pockets of the rock. And there are lots more complex shale oil pockets still to be drilled out. But remember, when you frack a tight oil deposit, it runs dry in about two years, and you have to move on to the next one. And the more money you have to spend getting the energy, the more the energy costs. But to hell with environmental consequences, let's frack till the wheels fall off!

Extremely high everything prices

Since literally everything we do requires oil (or some kind of energy to operate), everything will get more expensive. Food, gas, clothes, phones, EVERYTHING!

We've already talked a bit about how our miraculous world is built completely on the nearly free cost of shipping things overseas from one country to another in our mind-bogglingly complex supply chain. Obviously, this happens in massive container ships that have gotten so enormous, they're literally getting stuck in canals.

Never mind that 12 percent of global cargo runs through the Suez Canal, and that six-day blockage had knock-down effects for global supply chains when more than 400 ships were waiting to get through. And others still chose to go alllll the way around the tip of Africa instead of waiting. Oil prices spiked when 13 million barrels of oil were delayed . . . but never mind all that.

What do all these megaships run on? Oil. As we know, there is no battery powerful enough to move a ship of that size that far. Even gasoline isn't powerful enough. We need that thick stuff: diesel.¶ And we use so much of it, we don't even measure it in gallons. We measure it in "tons per hour." When the price of oil goes up, the price of shipping goes up. Remember when we were talking about all the raw materials it takes to make your iPhone and how we're gonna run out of those at some point? Well, even if we never run out of them, let's look at how many times those raw materials have to cross the planet in a shipping container to make it into your phone.

You probably have this idea that iPhones are made in China, but that's far from the case. While your iPhone was most likely assembled in Zhengzhou, China, by a company called Foxconn or Pegatron (which both happen to be Taiwanese companies), most of the parts that make your phone do what your phone does were built somewhere else. In fact, China isn't very good at making anything complex. But what they are good at is assembling things that other countries made. More than two hundred different companies spanning the entire globe supply Apple with the various parts that will eventually become your phone that's "Made in China."

So let's break it down. Your processor is almost definitely made in Taiwan (they are 100 percent the best at this) while the display is probably made in South Korea, and the Gorilla Glass that protects it might come from a Corning factory in the USA or Taiwan or Japan. Your battery was also made in Korea, except by Samsung. The flash memory was probably also from Japan (but a different factory, obviously). And the camera is made by Sony, which is also a Japanese company, but they have factories all over the world. The gyroscope that tells you where you are and what direction you're walking was probably made in Switzerland, while the accelerometer (which tells you how fast you're going) was made by German-based Bosch (but also could have been made at any of their global factories). The touchscreen and Wi-Fi chip are both "Made in the USA" . . . sort of. Because Broadcom and Murata

¶   To be more precise, 87% of marine shipping is powered by "Heavy Fuel Oil," or HFO. It's even thicker than diesel, and it burns way dirtier, but it's about 30% cheaper.

are US companies with factories in Japan, Brazil, and Thailand to name a few. And then it probably stopped by Indonesia for the soldering because they have lots of tin and that's something they're good at. And that's just the big stuff. There are also dozens of smaller components like USB microcontrollers, wireless chipsets, and OLED drivers that could have come from anywhere.

And all that shipping from two hundred different companies around the world is in addition to each of those two hundred companies getting

the dozens of raw materials they need from numerous other different countries in order to make the part that they finally ship to China.

It's a lot of fuel. But right now, it doesn't matter. Because it's so flippin' cheap. It's so cheap that it's cheaper to ship some pears 25,000 miles across entire planet—twice—than it is to just grow some pears and pack them in New York.

But increase the cost of oil by even 5 percent and it has massive ripple-through effects. Now imagine doubling or tripling or sextupling the cost of oil. And that's also on top of the geopolitical issues that are going to disrupt these incredibly fragile supply chains.**

Remember during the pandemic when the cost of trans-oceanic shipping containers increased by 700 percent? Well, the sea carries more than 80 percent of the world's traded goods. And when the cost of shipping increases by seven-fold, you see massive spikes in inflation.

The UK saw the price of "white goods" like washing machines and refrigerators increase 46 percent from January 2020 to February 2022. Prices for used cars inflated 37.3 percent in 2021 after a shortage of

---

** Once again, check out Peter Zeihan's book if you want some insight on how pirates might be making a comeback.

semiconductor chips brought new-car production to a halt. The list goes on and on. You were here; you remember!

Now, a lot of that inflation was companies taking advantage of global supply chain pressures to raise their prices anyway, just because they can. But either way you slice it, everything got a whole lot more expensive in 2021. And just because shipping container prices have dropped back to normal, doesn't mean companies will lower their prices. Because why on earth would they ever do that in a largely monopolistic system?

If we don't want to see massive disruptions in global supply chains, we have to discover a fuel that's powerful enough and cheap enough to move these massive container ships tens of thousands of miles across the ocean—dozens of times—just to make your phone. Or your blue jeans. Or the sewing machine that sewed those blue jeans in a totally different country than the cotton was grown. I really, really hope we figure out a way to electrify all of our transportation before oil starts getting insanely expensive. But it's far more likely that we don't.

I'm hungry.

You know what else depends on our magical world of dirt-cheap international shipping and global peace the likes of which the world has never seen before? Global food supply.

I want to give you a little challenge. Next time you're in the grocery store—once you're done marveling at the frozen pizza section—I want you to look at where each of the items in your cart comes from. Now, most of the regular boxed cereal or Tyson's chicken or whatever is produced in the good ol' U-S-of-A (though it may be from a state that's several thousand miles away). We make a LOT of our own food compared to many other countries. But those avocados? From Mexico. Your kiwis are from . . . well, New Zealand, obviously. Your coffee? Probably Colombia. Your wine? Probably France. Or Australia. Over 95 percent of coffee, cocoa, spices, and fish or shellfish products consumed in the United States are imported, as are about half of our fresh fruits and fruit juices, and almost a third of wine and sugar.[49]

And we're not alone. For the past fifty-odd years, countries have been able to cheaply import whatever food they need from wherever it's made. Meaning you could build up a population in a place that nature never could have previously supported. Look at the Middle East—they built entire modern megacities in the middle of literal deserts.†† They don't grow any of their own food. The only thing they grow is oil, which they export to have the money to buy food from people with arable land. But it's not just Saudi Arabia. Almost every country imports food from somewhere else. One in five calories in the entire world is traded, and one in six people around the entire world depends on imports to get the calories they need. Countries like Japan, as rich as they are, hardly make any of their own food. There are 34 countries in the world who are physically incapable today of producing their own food due to land and water constraints.

Now, the US is lucky in that we do most of our importing from two countries that are very close by: Canada and Mexico. But when the faraway countries we import stuff from start experiencing food insecurity, we will feel it too. And there will be a day when you're walking down those once-gloriously-stocked aisles and finding them . . . empty.

It's already happening. Remember the toilet paper crisis of 2020? It will be like that, but with basically everything. We're right in the middle of the "egg crisis of 2023." And if that toilet paper crisis (or the egg crisis) taught us anything, it's that Americans FREAK OUT when they think they're about to run out of something. And companies rampantly raise prices whenever they feel like they can get away with it. It probably won't surprise you to learn that profits at the largest US egg producer were up 718 percent for the first quarter of 2023. They didn't *have* to raise the prices of eggs; they just could, so they did. Because capitalism!

As the global population keeps growing, this is only going to get worse. What happens when an entire nation can no longer afford to import the foodstuffs they need to survive? Well, countries are going to have to make domestic food production a top priority to keep up with ever-expanding demand and stop producing . . . whatever else it is they produce.

---

†† I realize we did the exact same thing with Las Vegas, but their desert cities are wayyyy more ridiculous.

Back in 18th century England, before the food surplus was reliable enough to be sure the English people wouldn't starve, there were all kinds of restrictions on selling your wheat. Because back then, if some bad weather destroyed your crop, you'd fucking die. And the government was not about to sell wheat to another country if their own people might starve. It wasn't until they were reliably producing more food than they could eat every year that Adam Smith came in and was like, "Hey guys, I think we can actually make more money if we start trading more." And the government realized they weren't in danger of starving, and decided they could afford to sell their food to other countries (so they could also buy more stuff from other countries). They relaxed restrictions, lowered import taxes, and everybody got rich and fat! Hooray!

Now think about the opposite of that.

When a country has a shortage, they're going to close their doors and focus on taking care of their own people. And if they can't make enough food to feed their people, they certainly won't be exporting any of it. And the less food there is in the global supply chain, the more expensive it gets. The less *anything* there is in the global supply chain, the more expensive it gets. But it won't look that simple. As Zeihan explains:

> When things go haywire, it's not that everything gets more expensive everywhere. It's that complex inflationary and deflationary pressures in micro-economies around the world get real fucking confusing. We can't predict which thing is gonna do what. We only know that we won't know until it starts happening.

We may not be able to predict where all the prices end up based on which things we're running out of, but we do know that when we start having shortages, just like the toilet paper, people start acting weird. When those shortages are national, *countries* start acting weird. They look after their own needs first.

South Africa right now (and when I lived there) is going through controlled blackouts called "Load Shedding" lasting up to twelve hours a day in two-hour stints. You would know in advance what time the blackout was coming, and you'd just make sure your phone was charged, and get some candles ready, and then just sit in the relative dark for a few hours. Sometimes when they said it was two hours, it was really four. Apparently now, four-and-a-half hours is the regular. Their first move? Cut off any electricity they were exporting to neighboring Namibia, Zimbabwe, Mozambique, and Botswana of course, causing a ripple-effect through Southern Africa. If a country doesn't have enough electricity because they can't afford it, they can't make things or sell things—things that are all part of that wonderfully efficient global supply chain.

During load shedding, small businesses and grocery stores tried to stay open with generators, but most places just couldn't. Restaurants can't keep food cold, offices can't operate without internet, shops can't operate without their credit card payment terminals, and certainly factories can't do jack shit without the power on—it was bad. It's only gotten worse. And they were already running completely on coal. And they had plenty of it. This was just because their power stations were severely mismanaged and in disrepair. It doesn't take a stretch of the imagination to see this happening to plenty of other countries with weak power grids (or weak economies), causing them to lose the ability to manufacture much of anything at all. Sure, first they'll stop manufacturing baseball hats or whatever else to make as much food as possible, but if they're on that unlucky list of countries who physically can't produce enough food to feed their people . . . well, people will start dying. Lots of them. And fewer people means less chance of keeping anything that requires specialized labor working, whether it's road construction or the electrical grid or . . . food production. It's yet another very negative positive feedback loop.

And this is what decivilization looks like. It's an industrialized country sinking back into an agrarian economy because they don't have the energy to keep up their industry. Or they don't have the food. Or both. It's one thing breaking that breaks another thing because every single thing in our crazy complex world is dependent on all the other things. And then the ripple effects of that country decivilizing are felt in every country that relied on them—whether relying on money from them buying our exports, or relying on imports from them that we can't get anywhere else. It's one domino after another. It's absolutely bonkers how fragile this whole system is.

But maybe you're not worried about that scenario. Maybe "global manufacturing and food supply chains collapsing because people are dying of starvation in a gas-is-really-expensive-and-our-population-is-exploding world" isn't on your radar. But the food shortages are coming long before we run out of electricity or cheap oil. They're already here.

We're already seeing the tangible effects of a rapidly warming climate all over the world. Pakistan flooded into a massive lake, the American west is running out of water, Texas freezing, Portland melting, Southern Europe turning into a tinder box. As these "once-in-a-lifetime" weather events start happening every few months, crops will be destroyed. Food will be in short supply. Which means . . . you guessed it . . . it will be a LOT more expensive. And people will die.

Global food security is already in an exceptionally precarious place since Russia decided to invade Ukraine, blocking grain exports, and driving up fertilizer prices outside the reach of many farmers. The price of wheat in Africa is up 45 percent since before the war. The price of margarine is up 44 percent in America because we make it out of vegetable oil that comes from Ukraine. The floods in Pakistan are affecting global rice supply (also driving up prices), and the inflation we're experiencing in the US affects poor, debt-riddled countries even harder than us.

Why? Because inflation means rising interest rates. Which make food more expensive because the entire farming industry runs on debt. They take out loans for everything every season from seeds to fertilizer to machinery. So when inflation is high and the US government raises interest rates (to stop people from buying things in order to bring down demand, which supposedly lowers prices and fixes inflation, but doesn't unless the problem is "everyone has too much money"), then those borrowing costs for farmers increase. By a lot. The US farming sector's total interest expense is forecasted to hit nearly $26.5 billion in 2023—over 30 percent higher than 2022. And guess who pays the higher prices for corn or pasta or radishes? You do!

With all these different price pressures on global food supply including inflation, higher energy costs, and extremely high fertilizer costs, the United Nations estimates that in 2022, the global agricultural input import bill (the cost of energy, fertilizer, pesticides, and seeds to produce food that's going to be imported by someone else) is expected to jump to $424 billion, a 112 percent increase over 2020.[50] And the amount we're spending importing the actual food will reach $1.9 trillion worldwide, up 29.5 percent over the same period.

But the even crazier part is that these massive increases in the cost of imported foods aren't translating into any more food. Sub-Saharan Africa is going to spend $4.8 billion more on food imports for LESS actual food. While wealthy nations like the US and Germany and Japan and Saudi Arabia will, of course, keep paying whatever we have to pay to import the Sriracha and peanut butter and Vegemite and whatever other fun, exotic foods we like,[‡‡] middle-income countries like Vietnam and South Africa and Argentina won't be so lucky. Many of these countries will be forced to pare down to importing only staples they need to survive like rice and grain.[§§]

And then there's the reallllly unlucky countries like Ethiopia and Nigeria and Sudan and Yemen and Madagascar (the list goes on) that won't be able to import anything at all. And if your country can't make its own food and can't afford to import food, welp, people start starving. They're already starving. They've been starving for years.

In 2021, global hunger levels surpassed all previous records with nearly 924 million people (11.7 percent of the global population) facing food insecurity at severe levels, an increase of 207 million people in two years. Certainly a step in the wrong direction toward our lofty liberal-democratic goal of "ending world hunger by 2030."

---

‡‡   Peanut butter is pretty weird to the rest of the world.

§§   Still not worried about that whole decivilization spiral?

On top of that, a full 2.3 billion people in the world (nearly a third of the entire planet) were "moderately or severely food insecure" in 2021. That's 350 million more compared to before the outbreak of the COVID19 pandemic.

What's causing all this starvation? Well, war, conflict, inflation, rising prices, sure. But also, you guessed it: climate change!

Drought, rainfall deficits, flooding, and cyclones have all already contributed to crises in East, Central, and Southern Africa and Eurasia. And this is only just the beginning. Look out a little longer term, say 2080, and you could see the decimation of much of the world's rice crops due to rising sea levels that essentially salinate the low-lying lands off the coast, making them infertile. Often up to two hundred miles inland!

But maybe you're like, whatever, where even are those other countries? Who cares, I'm fine. The United States will always have food, so screw everybody else. Well, you're not wrong, Walter, you're just an asshole.

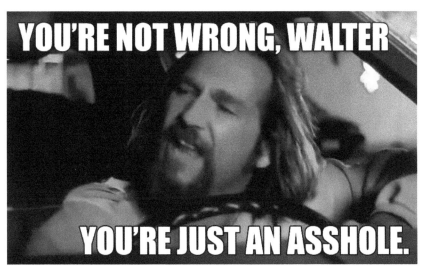

America is incredibly food secure compared to basically every other country in the world. We also have the benefit of having lots of money to use lots of new technology to increase crop yields. Things like even-further-genetically-modified seeds along with advances in "digital farming" where a machine goes around and checks each and every plant, could double crop yields AGAIN in the next ten years. Hooray! More jobs for robots! Just only for the countries who can afford this gigantic, very very expensive machinery and have row crops big enough that it makes sense (pretty much just the US, Canada, and Australia) and enough oil to power the machines and to pump that dead soil full of fertilizer. So you're not wrong, we can kick the can a little further down the road. Shit will be more expensive, but we're gonna be fine. Who cares if I only have ten choices in frozen pizza instead of 250? Who cares if a jar of pasta sauce costs $14? Higher prices are manageable in the richest country in the world, so long as we still have the food.

But what happens when the drought is so bad that you can't even grow the crops or raise the cattle where you once could?

According to the Farm Bureau Federation,[51] over 60 percent of the American West, Southwest, and Central Plains was categorized at "severe drought" or higher in 2021. These areas account for nearly half of all food production nationally. In the same year, nearly three-quarters of farmers saw a significant reduction in harvest yields due to drought, and many of them are being forced to fallow fields and sell the land to build . . . houses.

But houses for whom? Almost 40 million people in California, Arizona, Nevada, New Mexico, Utah, Colorado, and Wyoming already rely on the Colorado River for their water. Not to mention the twenty-nine Native American tribes, part of Mexico, and 5.5 million acres of farmland that need to be irrigated. All-told, the Colorado River basin spans about 246,000 square miles and represents eight percent of the land in the United States. And it's drying up the way Hemingway believed people went broke—gradually, then suddenly. The American Southwest is two decades into the worst drought they've seen in 1,200 years.[52]

These seven states have a complex water agreement that's been in place since 1922 called the Colorado River Compact that allocates all the flows of the river, with some states holding priority over others.* The problem is, even when they made the compact, they had already over-estimated the amount of water the river produced. They're currently using about 15 million acre-feet (the amount needed to cover an acre of land in a foot of water), but the historical flow is only about 12 million acre-feet.[53] And that was in a normal year. Now, according to climate research scientists at Colorado State University, the correct word is no longer "drought" but "aridification," since the conditions are unlikely to change. The biggest reservoirs (Lake Mead and Lake Powell created by the Hoover and Glen Canyon dams, respectively) have dropped 70 percent in the last twenty years. It also bears reminding that these dams provide nine billion kilowatt-hours of hydroelectricity annually, serving millions of people (and the entire city of Las Vegas that for some inexplicable reason they built in the middle of a desert and then filled with ostentatious fountains). No more water? No more power.

I hate to be Debbie Downer (I'm seriously no fun at parties), but it's only going to get worse.

We are definitely going to overshoot the 1.5 degrees of warming that lead to catastrophic, likely irreversible climate shifts, and it's just gonna be more floods, more drought, more fires, more hurricanes, more heat domes, more polar vortexes, and more uncertainty. Will we survive? Of course we will. Some of us, anyway. It just might get pretty bleak while millions (billions?) of people die of famine. Or, you live in Illinois where warming temperatures create more moisture, they're actually better for farming, they're seeing *increased* yields, and might even be able to plant a second round of crops when overnight frosts become a thing of the past! Let's all move to the Midwest!

Climate change isn't one-size-fits-all. But the places that are better suited to withstand it (or even benefit from it) are gonna have to watch a lot of places that aren't . . . basically collapse.

---

* Surprise! Mexico and the Native Americans were conveniently left out of this deal.

I'm thirsty.

Of course, water shortages don't just affect crops or hydroelectric dams. It turns out we also use water to . . . drink. And bathe. And cook. And wash dishes. Turns out we use a whole bunch of it every single day for basically everything. Turns out we die without it pretty fucking quickly. And while most of us in the United States probably feel like running out of water would just never happen . . . I beg you to look around. It's already happening in plenty of places around the world.

In South Africa, after a drought that only lasted about three years, the city of Cape Town allllllllmost ran out of water. There were countdown clocks to "Day Zero" when the city's reservoirs basically turned to sludge. That's right, in the three years I lived there, we were running out of both water *and* power. The city had already created a plan for when the taps ran dry that involved lining up at water collection points under the watch of armed guards to collect a measly six gallons of water per person. And in case you're not familiar with Cape Town, it's a gorgeous metropolitan city of four million people. Not some African village made of mud huts. Its pristine beaches are speckled with multi-million-dollar mansions, and the city is home to two of the world's fifty best restaurants.

I lived in Cape Town from 2017 to 2020, and for much of that time we had to take ninety-second showers (turning the water off in between soaping and rinsing) while standing in a bucket, and then use that bucket water to flush the toilet, which you would only do once a day. No one let the water run while brushing their teeth or washing dishes in the sink. Every yellow was left to mellow—even in public spaces— and bathrooms in restaurants starting reeking of ammonia. So, products were developed to help lessen the stench of old urine sitting in toilets. Every single drop was saved. Except for those rich assholes who broke all the rules and kept filling their pools and watering their green grass. And you can bet that grocery store shelves were immediately emptied of any water delivery they got as people got in line before opening hours to stock up. I tried every day for months, and never got a single jug. Rich people installed giant cisterns and grey water toilets, and poor people waited in line for hours at natural mountain springs to fill up a jug or two.

As it neared Day Zero, and we all prayed for rain, each person was allowed only 13 gallons of water per day. Not sure how much 13 gallons is? Well,

A full bath? 30 gallons
A ten-minute shower: 25 gallons
Toilet flushing: 1.6 gallons per flush
Washing machine load: 20 gallons
Dishwasher: 4–10 gallons
Drinking water: 1 gallon
Hygiene: About 3 gallons

So the water we were each allocated added up to a ninety-second shower, a half-gallon of drinking water, a sink-full to hand-wash dishes or laundry, one cooked meal, two hand washings, two teeth brushings, and one toilet flush *for the house to share*. We never hit Day Zero. But that's only because the rainy season came—the one that had not produced nearly enough rain for the past three years—and it finally fucking rained. And because the people of Cape Town managed to cut their water consumption by more than 50 percent. We avoided Day Zero through personal sacrifice and a stroke of luck. The dams slowly refilled, and we breathed a collective sigh of relief. No new technology saved us. It was just the weather. And plenty of places on Earth won't be so lucky. And let me be clear: NO ONE in Cape Town went back to letting the water run while brushing their teeth or doing the dishes. Once you realize how truly life-giving and precious it is, you'll never waste it again.[†]

Mexico is in the midst of a similar crisis . . . and it's been going on for years. Several cities across the country have already reached the "Day Zero" that Cape Town so narrowly avoided. Nearly two-thirds of

---

† Then I came back to the States and watched my mother turn on the faucet in the kitchen, only to turn and walk away, leaving it running for entire minutes! I think I have PTSD from my year under the threat of Day Zero.

the country's municipalities are facing some level of shortage. Brown, brackish water that isn't even suitable for drinking is delivered by trucks called *pipas*. And of course people are hijacking those trucks to steal the dirty water in them before it makes its deliveries. Then you have to pay jacked up prices to the water thieves, obviously. If your *pipa* makes it, you can expect to line up with as many buckets as you can carry to take your weekly allocation home. And naturally, even buckets are in short supply and now cost three times what they used to because capitalism, amirite?

Or, if you're lucky enough to be able to afford it, you can buy bottled water in a store, but it now costs more than gasoline. In Monterrey, the worst-hit region, their reservoir is completely dry. It's sand. You can walk across the bottom of it. In another fun twist, places like Mexico City have been taking water from aquifers underneath the city since forever ago, which, since they're not replenished by rain, get lower and lower, and now the city is sinking. Infrastructure crumbles, buildings topple, it's super fun!

Okay, so South Africa and Mexico . . . not exactly the most industrialized countries out there. It's not like that would ever happen in the US! Except, it already is. An Arizona suburb just sued the town of Scottsdale in January 2023 for cutting off water to their small, unincorporated town due to the severity of the drought. They're skipping showers and eating off paper plates while they try to find someone to truck water in, which obviously will cost a whole lot more money. The residents of this wealthy suburb full of $500,000 homes at least have the money to pay no matter how much it costs.

But whatever, there will always be plenty of water so long as you've got that cash! Well, maybe. According to the Global Commission on the Economics of Water,[54] global water demand is expected to outstrip supply by the end of THIS decade. Not 2040 or 2050 or 2080. This one. The 2020s. And not just by one or two percent, but by 40 fucking percent. Some countries like Saudi Arabia are already using 35× the amount of water their landscapes can naturally support. What did we say about building cities in the middle of the desert???

Though I guess we aren't really ones to talk considering we have perfectly clean water . . . and we shit in it.

Let's get the heck outta Dodge.

Now that some places are running out of food, and some places are running out of fresh water, what do you think people will do? Just hang out and die? Of course not! We're going to migrate—as we've always done for all of human history—to find a place where we're less likely to die. It's just what we do.

And this isn't just something that might happen, it's something that's *already* happening. According to the International Displacement Monitoring Centre, weather-related disasters force an average of 21.8 million people to flee their homes every year. While a lot of these people return when the hurricane is over or the fire gets put out, there are going to be more and more places that you just can't return to.

The foreign minister of the tiny island nation of Tuvalu recently made a speech standing knee-deep in seawater (that used to be land) to show that his country is literally being swallowed by rising sea levels. At least eight smaller Pacific islands have already *completely disappeared*, and the Marshall Islands are on track to get submerged as well. The National Oceanic and Atmospheric Administration (NOAA) estimates sea levels could rise up to *seven feet* by the end of the century if we fail to curb future emissions.‡ But it doesn't have to swallow entire cities to make them basically uninhabitable.

In the United States, nearly 30 percent of the population lives in high-population-density, low-lying coastal areas. These areas will continue to be hit by rising sea levels full of warmer water driving harder, more frequent, more destructive storm surges that reach ever further inland. Rising seas threaten infrastructure and local industry—roads, bridges, subways, oil and gas wells, power plants, sewage treatment plants, landfills—the list is endless. And obviously we need all those sort-of-gross things we don't like to think about functioning in order to keep the Incredible Everything Machine operating in its pleasantly invisible manner.

Recent torrential rain causing unprecedented flooding in Jackson, Mississippi (coupled with years of poorly managed water treatment facilities) led to the failure of those treatment plants, forcing residents to line up in cars for hours for cases of bottled water—only to be turned away when the city didn't have enough. I don't think we need to talk about Flint, Michigan.

According to a 2021 report by the World Bank,[55] climate change could force more than 216 million people from their homes by 2050 and hotspots could start emerging by 2030. Sure, that's only like 3 percent of the global population, but it will feel a lot bigger when they start moving into your town. Two-thirds of Americans already think we need to stop or reduce immigration.[56] Something tells me we're not gonna be shouting "give me your tired, your poor, your huddled masses" when the rest of Central America wants to move north.

But let's get a little more specific about what could possibly cause someone to abandon the only home they've ever known. Obviously, if your country is now underwater or has no water, you don't have much of

---

‡　Here is a fun calculator you can play with to see how different volumes of sea-level rise will affect global coastlines: https://coast.noaa.gov/slr/#/layer/slr/7/-10957408.272357693/4740085.158735173/5/satellite/none/0.8/2050/interHigh/midAccretion

a choice. But it will also be, once again, like Hemingway's going broke: gradually, then suddenly.

People will leave areas with lower water availability and crop productivity, as well as places where the sea level has risen (even the ones that aren't fully underwater), and places where the storms are just too intense too often. If you used to be able to grow rice, and now you can't grow rice because the land is too wet (or too dry, rice is very finicky), you're gonna leave. People will internally migrate in the US as the Southwest turns into a desert and water gets harder to come by. People will leave Arizona if it turns into Death Valley.§

In Southeast Asia (Cambodia, Thailand, Vietnam, Indonesia, Laos, Malaysia, the Philippines, Singapore, Myanmar, Brunei, and Timor-Leste), climate change is expected to hit the hardest, as typhoons and flooding increase in intensity each year, and much of the nearly 700 million people who live there live in low-lying coastal areas.

Warmer weather means stronger, more frequent storms, means you can't grow as much rice as you used to, and your house keeps getting destroyed, and also probably some of your friends have died. Or you lost your cattle. Or all of the above. According to the IMF, in the absence of new technology, rice yields in Indonesia, the Philippines, Thailand, and Vietnam could drop by as much as 50 percent by 2100 (as compared to levels from 1990). That's not great when your population is still rapidly expanding. And everybody eats rice for every meal. And since their populations and economies are both growing, it means they're using a whole lot more energy. Overall energy demand in Southeast Asia is expected to nearly double by 2050.[57] That's basically the equivalent of adding an extra hundred million humans living at American levels of consumption. Oh, and I'll give you one guess as to what they're burning to get all that energy.¶

Even more fun is Southeast Asia's energy needs are set to outpace their available supply (even of dirty, dirty coal), which means they'll have to start importing it. Which you need money to do. Except, according to

---

§   Have you noticed how nobody lives in Death Valley?

¶   If you said "coal," congratulations! You win! But everybody loses.

the Asian Development Bank, climate change could shave 11 percent off the region's GDP by the end of the century, forcing millions of people back into poverty . . . soooooo anyway, things aren't looking great.

And just in case you really don't care about nearly three-quarters-of-a-billion people who live on the other side of the world (and who happened to have invented basically all of the most delicious food in the world), where do you think so many of those wonderfully cheap products we love to buy get made? You guessed it. Your T-shirt hasn't said "Made in China" for a while. China got expensive. Try Malaysia and Vietnam.

Okay, so let's say we invent some super cool, genetically modified, flood-resistant rice (versions already exist, and they'll surely get better) and everybody decides to stay despite their poorly constructed houses getting razed by a typhoon every other year. There's another problem that's expected to affect this region (as well as a whole bunch of others around the globe). And that's:

Wet-bulb temperature! What the heck is a wet-bulb temperature, you ask? The wet-bulb temperature (WBT) is measured by covering a thermometer in a water-soaked cloth. At 100 percent humidity, the WBT is equal to the air temperature. The lower the humidity gets, the lower the WBT gets, as it's cooled by evaporation. This is also how human bodies work.

Why are we talking about this? Well, it's been in the news a bit recently. Most notably, after the deadly Pacific Northwest heat wave of 2021, John Oliver did a segment on his show about the lack of air conditioning in prisons and public schools across the country. Why? Because 120°F (49°C) is painfully sweltering, sure. But as long as the air isn't too humid, and you have enough water, your body will sweat, and the sweat will evaporate, and you'll cool down. You won't be comfortable, but you won't die. However, there is a tipping point of high heat and high humidity where the body can no longer cool itself. The air is too full of water already, so it can't absorb the sweat on your body, and you just . . . cook. It probably goes without saying, but—just to be clear—when you cook, you die.

Right now, there aren't any places on the planet where people regularly just cook to death, but it happens. It happened when the Pacific Northwest was stuck in that "heat dome" for a week; it happens when prisons and schools in places that didn't use to need air conditioning don't have it, and as the world continues to warm, it will happen in more and more humid places that just didn't use to get so effing hot. A fun side effect is more people using air conditioning means using more energy that burns more fossil fuels, which only further exacerbates the problem! A dangerous WBT doesn't necessarily have to mean it gets hotter, either. It could stay the same temperature but just get way more humid, which is something climate change is also likely to do in certain pockets around the world.

The WBT considered to be fatal is 35°C (95°F). Now, plenty of places get above that temperature all the time, but as they say, *it's not the heat, it's the humidity*. In the 2003 European and 2010 Russian heat waves—that killed an estimated 20,000 and 56,000 people, respectively—people were dying from WBTs of just 28°C (82.4°F). Especially if you're old, or you're working outside, or you aren't used to extreme heat, so you don't know how to react . . . but *everybody* dies at 35°C (95°F). Even if you're sitting in the shade with an unlimited water supply.

So what does all this have to do with migration? Well, duh. As we continue to see more and more places reaching fatal wet-bulb temperatures more regularly, people are gonna get the fuck out (or die trying). And while climate scientists can't predict exactly where this is going to happen, NASA's climate science division says the most vulnerable places on the planet are South Asia (this includes the entire billion people living in India plus another nine countries around it), the Persian Gulf and the Red Sea (possibly in the next thirty years), and then Eastern China, parts of Southeast Asia, and Brazil (more like the next fifty years).[58] And don't think they're leaving out America! Midwestern states like Arkansas, Missouri, and Iowa are the highest risk and will likely hit the critical limit in the next fifty years as well.

But even before we all start just dropping dead from WBTs, things are going to keep getting worse. According to the First Street Foundation, an "Extreme Heat Belt" will impact over 107 million Americans by 2053.[59] That's only thirty years away, and that's almost a third of our national population. In 2023, just fifty counties in the US are expected to see temperatures over 125°F. By 2053, that number is projected to break a thousand. Even if you're not dying from WBTs, this extreme heat belt means more and more days of unbearable heat each year for millions more people. Unbearable heat that makes people think about migrating to a more agreeable place.

Oh, and there's one other fun, temperature-related thing at play here: the temperature at which mammals can no longer reproduce. And that number is much lower than the temperature we just die at. If you're a dude, have you ever wondered why your ball sack scrunches up when it's cold and hangs down like your grandpa's jowls when it's hot? It's because sperm is super-duper sensitive to temperature variation and needs to stay a few degrees cooler than the rest of your body. So our bodies (well, your bodies) have a special muscle called the cremaster muscle that works to keep your sperm at the perfect temperature for baby making. You've got a whole Goldilocks thing going on down there.

And what's true for us is also true for the mammals we eat. In order to raise your goats and your cows, they have to be able to make more goats and cows. But above certain temperatures, that's just not gonna happen. What's the exact temperature? We're not sure. One study suggested the upper limit for optimal reproduction is 25–26°C (77–79°F).[60] But what we do know is the hotter it gets, the less likely mammals are to reproduce. And if your entire existence is dependent on cattle, and you live in an area that's getting super hot (like, say, the Masai tribes of East Africa), you're gonna have to figure something else out when the cows stop making more cows.

Hey, let's just take that guy's food!

You know what else causes mass migration? War! You know what causes a lot of wars? Resource scarcity! When food gets scarce, when energy gets scarce, countries get all Lord-of-the-Flies-y, every-man-for-himself, and start taking whatever they can get their hands on. You think it's a coincidence that Russia invaded Ukraine when more than 55 percent of Ukrainian land is arable, they are the fifth-largest wheat exporter, and they account for more than 40 percent of global sunflower seed oil production?

Russia wants that damn food (and the oil, obvi).

Now let's look at some other wars either caused by drought or famine, or caus*ing* famine 'cause it can go both ways, which makes it super fun.

You know about that whole war in Syria, right? No? Oh, well, back in 2011, civil war broke out where more than 300,000 people have been killed and 6.6 million Syrians (in a country of 17 million people) were displaced from their homes, nbd. Now, this was due in part to extreme political instability leading to a violent regime change and some other complex geopolitical factors, but a lot of analysts also think a massive drought in the country played a not-so-insignificant part. Syria's drought lasted from 2006 to 2010 and was recorded as the worst multi-year drought in around 900 years.

Less rain along with rising temperatures led to desertification of agricultural land, which led to 800,000 people losing their income, and 85 percent of the country's livestock dying. Crop yields plummeted by up to two-thirds, and the country had to start importing food, which made it way more expensive. With all this going on, over a million rural workers headed into the cities for work. These rural workers were extremely poor and *still* couldn't afford the more expensive food, which made them ideal targets for recruitment by terrorist groups like the Islamic State.[61] Add in reduced food and fuel subsidies in an incredibly unstable region, and you get massive unrest and a recipe for disaster. Just to be clear, this war is still happening. According to Aron Lund, a Middle East analyst at the Swedish Defense Research Agency:

There are crippling shortages of key imports, energy and water. New UN data says 15.3 million Syrians now depend on humanitarian assistance, or nearly 70 percent of the country's current population. [. . .] Even though violence has ebbed to its lowest point, the situation for civilians is, paradoxically, worse than at any previous time.[62]

Now, there are also plenty of analysts who deny that climate change had anything to do with anything, that the numbers of people displaced due to drought are exaggerated, that the drought didn't even really happen, and everybody just needs to chill out.[63] But even if the Syrian civil war wasn't exacerbated by drought and famine, we can use this example as a cautionary tale of what *can* happen when millions of people in a country suddenly don't have access to food and water. Governments freak out, people get crazy, and just maybe you have a civil war on your hands.

Speaking of civil war, let's talk about the other side: when it's the war that causes the famine. The Middle East country of Yemen has been dying of starvation *en masse* since 2016. But this didn't happen because of a drought or flood or crops dying. It happened because they're in the midst of a civil war. Saudi Arabia decided to get involved, created a blockade to stop any food from getting in, bombed farms, destroyed all the boats people used to fish—literally specifically, intentionally starving the country to death. And it'll make you feel warm and fuzzy inside to know the US is supplying Saudi Arabia with the weapons and intelligence to make it happen.

When resources get scarce, countries will go to war. And when countries go to war, and people get desperate, people don't just die from bombs. They starve. They are starved as a tactic for winning a war. Wait, doesn't the Geneva Convention say something about that?

This section could be a whole lot longer. There are whole books on subsections of this section. But these are just a few examples of things that are already happening (that didn't used to happen so much) that could start inspiring tens or hundreds of millions of people to get the heck outta Dodge. Will the planet survive people moving all over the place? Sure! Will all the people survive? Definitely not.

The thing is, these problems probably don't seem that new or novel. There have always been commercials asking rich Americans to help save the starving children in Ethiopia or Somalia or Yemen. People have always been starving, there have always been wars, so what's the difference? Well, the difference is you probably haven't experienced a war or a famine your whole life. An isolated pocket of hungry people in Syria is not what we're talking about here. We're talking about massive, global famines caused by acute and long-term resource scarcity that mostly only wealthy, well-endowed countries will be able to navigate without losing an upsetting chunk of their population.

And just to close out this short-term section with one more tiny reminder: even if you don't really care about millions of people dying all over the world, you will still feel it. The shelves will be empty, and the food you can buy will cost twice as much. People in the United States will starve. In a country with limited social safety nets, stagnant wages, and already skyrocketing food prices, people are already choosing between heat and food in the richest country in the world. And if gas prices increase by 700 percent and food prices increase by 500 percent but your boss doesn't give you a 2,000 percent raise, you might be one of those people too. And even if you can still afford the food and heat and basic necessities, we all need to start imagining a world without that abundant, cheap energy that basically everything depends on. A world that looks different. That feels harder. A world where nothing is as easy as it is today, because this is the easiest it's ever going to be in the entire span of human history. And finally, if you're only worried about whether or not you'll personally survive the upcoming tumult—maybe it's time to stop being such an asshole.

The Long Term

**Posterity doesn't vote, and doesn't exert much
influence in the marketplace. So the living
go on stealing from their descendants.**

—William Catton Jr.
Overshoot

Well, that was fun! Now let's talk about the real fun stuff. Climate change is changing . . . everything. And while the planet is likely to survive with or without humanity, things might just look a little . . . different. To be clear, I'm not saying all these things are definitely going to happen; they're just definitely going to happen if we don't do something to stop them from happening. If you don't think climate change is "real" then you probably haven't gotten this far in the book anyway (or ever picked it up in the first place), but I highly recommend you put the book down now if you fall into the "global warming is a hoax" category. You clearly aren't swayed by science or facts or things your own eyeballs can see.

Farewell, weather as we know it.

You probably don't think about currents much in your daily life. But you should. Because they literally dictate the weather wherever you live. Currents are the moving of the water and air around the planet in ways that, up till now, have been very predictable. Here's a happy little map illustrating all the ocean currents we have and which ways they always move.

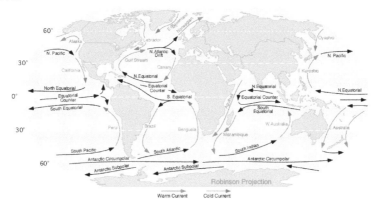

The ocean currents on our planet act like a conveyor belt, bringing warm water and precipitation from the equator out to the poles, and cold water back into the tropics. This levels out what would otherwise be pretty extreme temperature variations between the two (and they're already pretty extreme from a human perspective). The ocean near the tropics absorbs the vast majority of the radiation from the Sun that hits the Earth each year. Like, way more than the land. That water gets real warm. But luckily, we have ocean currents that distribute that warm water around, which regulates and stabilizes climate patterns around the globe.

But what causes the ocean to move in the first place? The currents are caused by a few things. First, there's the shallow currents close to the surface that are mostly caused by wind. They're also partially caused by temperature differentials in the water itself, along with the Earth's rotation, but it's mostly the wind that's doing the moving. Temperature differentials between the poles and the equator create wind (when high-pressure air moves toward low-pressure air), and wind moves the water in very predictable ways. We call these "prevailing winds." The bigger the temperature difference, the stronger the winds, the stronger the currents are.

Then there are deep ocean currents. These are mostly driven by differences in density. The colder and saltier the ocean water is, the denser it is. So the cold, dense water sinks, and the warm, less-dense water floats up. Except our surface ocean currents do this cool thing where they push down the warmer water where it becomes more dense, and then sinks, which pushes back up the cold, nutrient-dense water. The bigger the density differences in all these crazy layers of water, the more movement there is, and the more stable the climate is overall. We like movement. The more movement the better. All of the water in our oceans is constantly moving up, down, left, and right, circulating the globe in a thousand-year cycle. The volume of water transported by this global current conveyor belt is equal to 100 Amazon Rivers or 16× the flow of all the world's rivers combined. That's a lot of movement. And if we didn't have this movement, our planet would be a lot less habitable. The equator would be way too hot, and most everywhere else would be way too cold.

Remember that movie *The Day After Tomorrow* where the whole climate gets destabilized, and the entire Northern Hemisphere gets covered in an ice sheet? And then Dennis Quaid walks ninety miles in the sub-arctic temperatures to save his son, Jake Gyllenhaal, from the New York Public Library that's now buried under fifty feet of snow? And of course, a pack of wolves gets loose, because why not? That part's not going to happen. But in the movie, the stopping of the ocean currents is what causes the world to shift into an ice age overnight.

For a more practical example of how this works, we can talk about the Gulf Stream. You've probably heard of it. The Gulf Steam carries warm water up from the Gulf of Mexico near Florida all the way to the Norwegian Sea. The warm surface water becomes cooler, saltier, and denser as it moves north along the East Coast of the US (the Atlantic just happens to be extra salty), before it gets too salty and heavy and cold, sinks to the bottom, and flows along the bottom of the ocean all the way down to the equator, where the whole conveyor belt starts again. But on its journey north, this warmer water passes by the UK and mainland Europe, which is why the whole continent is much warmer than North American land that exists at the same exact latitude. It's estimated that Northern Europe is up to 9°C warmer (16.2°F)[64] because of the Gulf

Stream. Cities like London, Paris, Amsterdam, and Berlin would all be covered in several feet of snow every winter if they weren't blessed by that warm Gulf Stream air and water. Not only that, the Gulf Stream is such a fast and reliable route to England that Ben Franklin (and lots of other sailors) followed it to reduce their transatlantic transit time. The warmer water flows at an average speed of 6.4 km per hour (about 4 mph, or 3.5 knots). Considering those big wooden ships back in the day didn't travel much faster than 4 knots on their own, this was a huge bonus. Here is Ben Franklin's Gulf Stream map, just for fun.

Okay, so we know what the Gulf Stream is, and we like it. We LOVE it. But now for the sad part. Because of the way global warming works, the poles will always heat up much faster than the equator.** While the rest of the Earth has warmed about 0.8°C (1.44°F) in the last forty years, the Arctic has warmed by more than 3.5°F in the same period.[65] Yikes. The reason this matters is because when the poles get hotter and hotter and the equator stays basically the same, we get a smaller temperature differential between the two. And temperature differentials are what create wind. And wind is what drives the surface currents in the oceans . . . you see where I'm going with this? Not to

---

** We're still not 100 percent sure *why*, we just know it's true.

mention that renewable power source we've been betting hundreds of billions of dollars on . . .

To add insult to injury, warming at the poles also means melting ice caps. When ice caps melt, it doesn't just reduce the albedo and speed up warming, it also pours a whole bunch of fresh water into the ocean. More freshwater in the ocean means less salty water. And when you reduce salinity, the water doesn't get as dense, so it doesn't sink, which also affects the conveyor belt's ability to . . . convey. Triple whammy.

With the warming and the de-winding, the Gulf Stream (known in scientific circles as the Atlantic Meridional Overturning Circulation, or AMOC) has already slowed about 15 percent since the mid-20th century and is now at its weakest point in 1,600 years.[66] Is it going to stop? Maybe? Probably? This isn't gonna happen next year or anything, but think like, next 50 to 100 years? Or maybe even 200 years? It's not really easy to predict. We can't even really tell if or how much the winds are slowing down. The Earth is a super complex place. But even smaller amounts of slowing in the AMOC can have massive impacts on the global climate.

I know what you're thinking: "So what, Berlin is gonna be cold AF. I thought this was global *warm*ing!" But ocean currents have another really, really important function—arguably more important than the general habitability of our entire planet. *All* life in the ocean depends on these currents to oxygenate the water. Remember two minutes ago when we talked about "the more movement the better?" Well, that's true for oxygenation as well. The more the ocean churns and flips and mixes it up, the more oxygen exists in the water. And the colder water gets, the more oxygen it can hold. So when that Gulf Stream water reaches its maximum coldness and saltiness, it's also filled with oxygen. And then somewhere between Greenland and Iceland, it sinks to the bottom of the ocean and moves south, bringing oxygen to all the weird sci-fi creatures that live down there (as well as all the normal ones that feed all the other things).

What happens when we stop the conveyor belt? Well, when the ocean stops moving and flipping, it becomes stagnant. You know what stagnant water is? A swamp. According to Peter Ward, author of more than twenty books on the topic, you get what's called a Canfield Ocean.

Which is basically just a fancy term for swamp. Stagnant, swampy oceans existed during the Paleocene-Eocene Thermal Maximum about 55 million years ago (as well as plenty of other times). During the PETM, global temperatures increased about 5–8°C due to massive amounts of carbon dioxide being released into the atmosphere. Of course, 55 million years ago, it wasn't caused by humans driving cars, but by Mother Nature ripping a big one in the form of volcanic activity. So we're essentially just blowing up massive volcanoes every single day all over the planet. The last transition to a Canfield Ocean took a few hundred thousand to a million years. We're on track to do it in a few hundred.

Now, 5–8°C may not sound like a lot, but at those levels, NO ice existed on the whole planet. Sea level was hundreds of meters higher than it is now. And the ocean was a disgusting, deoxygenated swamp. And when that happens, unsurprisingly, lots of things start dying.

But it's not just the higher temperatures that kill off all the humans (or whatever was living 55 million years ago). It's a poisonous gas that swamps release when they're filled with billions of dying sea creatures! When things die, they decompose. And when they decompose in oxygen-starved environments, they release a toxic gas called hydrogen sulfide that kills . . . everything.

But on the bright side, it's possible we don't even have enough fossil fuels to release enough CO_2 to get us back to the toxic, iceless swamp planet, so at least we've got that going for us.

Of course, even if we don't go full Canfield Ocean, warming, deoxygenated oceans are still a problem. That toxic gas is still a problem.

Hydrogen sulfide already exists all over the planet in places where large amounts of organic matter tried to decompose in a place with very little oxygen. Like when we drill for oil, we have to check to make sure massive pockets of hydrogen sulfide aren't going to pop out and kill the oil workers. And sometimes they do. In case you're wondering, hydrogen sulfide smells like rotten eggs. So if you're on a beach covered in seaweed that smells like rotten eggs? Maybe get the fuck out.

Here's a place where it's already happening: Puget Sound in Seattle used to be surrounded by old-growth conifers (evergreens) that don't shed their leaves. But those got cut down over the years and have been replaced by deciduous trees that shed their leaves every fall. In addition, organic wood waste from sawmills gets dumped into the sound.[67] So now the sea floor is covered in leaves and sawdust, which kills the seagrass underneath. The seagrass provides a vital habitat for baby salmon and other marine organisms. But not only does the death of seagrass in an oxygen-starved environment release hydrogen sulfide and kill baby salmon and destroy the habitat of the baby salmon, the seagrass itself is an incredibly important carbon sink for the planet. It's like a rainforest in the ocean, sucking up CO_2 out of the water and releasing oxygen. So it's a vicious cycle of deoxygenation.

And this is just one little pocket we're creating in one little corner of the world. The less oxygen the oceans have, the more hydrogen sulfide we get. Warmer water holds less oxygen, and all of our water is getting warmer. And the cycle continues.

In 2011 there were around 700 reported sites worldwide affected by low oxygen conditions, up from only 45 before the 1960s. The volume of anoxic ocean waters—areas completely depleted of oxygen—has quadrupled since the 1960s.[68]

Overall, oxygen in the ocean has dropped about 2 percent in the last forty years, and we'll probably see another 3–4 percent drop by the end of the century, though it will be concentrated in different pockets around the world. Don't think 3–4 percent sounds like a lot? Well, most of that oxygen loss is concentrated in the first thousand meters of depth—where almost all of the animals live. And you better believe if you changed the oxygen concentration of the air we breathe by four percent, we'd notice it. Let me assure you, the fish are noticing it.

Farewell, biodiversity.

Have you ever heard of mass extinction events? A mass extinction event is a short period of time (geologically speaking) where a high percentage of all the species on earth—bacteria, fungi, plants, fish, whatever—dies out. The magic number is 75 percent of all *species* on earth. You're probably most familiar with the dinosaurs. But you probably haven't heard . . . the next one has already begun. We're smack dab in the middle of it. Either the sixth or the ninth mass extinction, depending on who you ask.

Usually, mass extinctions are caused by massive volcano eruptions, which release a bunch of CO_2 and change the whole vibe of the planet. Canfield Oceans and all that jazz. And then of course there was the asteroid that killed the dinos.

But this time around, it's a little different. Because for the first time in all of geological history, we're the ones doing it. And we *know* we're doing it, and we're still fucking doing it. While it's still debated a bit in science (like most things), 70 percent of biologists acknowledge this ongoing anthropogenic (human-caused) extinction event. And if you think about it, it seems kind of obvious.

Right now, about 40 percent of land globally has been converted for food production. Agriculture is responsible for 90 percent of global deforestation and 70 percent of freshwater use. The Amazon isn't burning because they can't put it out—they're setting it on fire to raise cattle or grow food for cattle. It's not a hyper complicated chain of events we're looking at. (I mean it is, but sometimes it isn't.) We're destroying the home where most of the species live. Just like if your island nation gets submerged in the Pacific Ocean, you gotta find a new place to live. And if you can't, you die. The same thing that's happening to the people of

Tuvalu is happening to every species in every biome we clearcut to make more room for industrial agriculture to feed all these people we keep creating. And then, not only does the clearcutting destroy the habitat for all the bugs and birds and bees, it also takes a carbon sink and turns it into a carbon source (both from the deforestation and the fossil-fuel intensity of industrial agriculture).

And of course, that whole global warming thing science nerds won't shut up about isn't just making the planet extra hurricane-friendly. Turns out coral—that supports 25 percent of all life in the ocean—is super sensitive. And since the ocean absorbs the vast majority of heat from the sun, it's warming up extra fast. And when the corals die, the fish die, when the fish die, the things that eat the fish die . . . you get the picture. Oh, plus that whole ocean deoxygenation thing we just talked about. Reaching 1.5 degrees of warming is predicted to kill 70 to 90 percent of all the coral in all the oceans in all the world. Two degrees? We're looking at 99 percent.[69] Is nature incredible and will some corals better adapted to warmer waters start popping up? Of course! Will the whole ocean ecosystem that 10 to 12 percent of the world population depends on for food and employment collapse before evolution kicks in? It sure seems like a real possibility!

And of course, it's not just the birds and the fish. We're also at critical levels for insect populations. And this is the weirdest one, because besides habitat destruction, we've just confused them with our lights. If you've ever turned on a light on your porch at night, it's no secret that bugs like it. Well, that's because lots of bugs use light to tell them when and where to have sex. But with all of our artificial lights, the bugs are too confused and don't know where to have sex so instead they die alone on your porch. Poor guy. He was just trying to get laid. Take a look at your windshield next time you're taking a road trip. You'll notice a suspicious absence of those things you used to have squeegee off every hundred miles.

Okay, so things die, but that's natural. Humans rule the world, we can do what we want. We're here, we're murderers, get used to it. But this is not an average amount of species dying. Currently, the species extinction rate is about 1,000 to 10,000 times higher than it's supposed to be. But even crazier is that our current extinction rate is 10 to 100 times higher than all the other mass extinctions! We're even outdoing the asteroid that killed the dinosaurs.

Remember when Portland melted back in the summer of 2021? And had temperatures higher than the record highs of fucking LAS VEGAS? Well, unsurprisingly that warmed up the shallow waters pretty significantly, and in the process killed BILLIONS of sea creatures. In a week. Not to mention all the humans who died in a place where only 30 percent of buildings have air conditioning—because they've never needed it before. But no big deal, I'm sure the death of billions of marine creatures in a very small area had no effect on anything else around it. So, we can't eat oysters for a bit. We'll survive.

Farewell, entire ecosystem and life as we know it.

I hate to break it to you, but the pesky thing about the ecosystem is everything is dependent on everything. You take out one thing and the thing that ate that thing has nothing to live on, and then the thing that ate that thing dies . . . yada yada yada . . . everybody's dead.

Tracey Thorn ✓
@tracey_thorn · Follow

I remember when I used to see a bee and go, YIKES a bee! And now I'm all, Oh wow a bee, hi! You ok there? Need anything? Can I get you a drink? A cushion? Wanna borrow the car?

5:03 PM · May 23, 2019

❤ 46.2K   Reply   Copy link

Read 413 replies

If we don't have bees to pollinate our plants, we don't have a way to pollinate them, except by hand, which is as tedious and time consuming as it sounds.

But of course, no bees doesn't just mean "no honey." It means no blueberries or cherries that rely on bees for 90 percent of their pollination. It means the entire sub-species of "bee-eater" bird (some of the most beautiful birds in the world) have nothing to eat. According to one source, 87 major crops worldwide employ animal pollinators (largely bees), and only 28 can survive without assistance from animal pollinators. If we lose the wasps too, you can say goodbye to figs forever.

Can we innovate our way out of pollinators like bees? I mean maybe, but JFC. I cannot even begin to fathom what that completely robotic, nature-less world would be like.

Either way, humans would survive the great bee extinction. We mostly survive on cereals that get pollinated by wind. But the ecosystem is a strange and delicate place. In Canada, acid rain is causing lakes to turn into literal jelly. Because the acid leaches calcium from the clay in the lakebed, and the organisms that depend on that calcium die, and they're getting replaced by these other organisms that just happen to be covered in jelly as a defense mechanism. They only need about 10 percent as much calcium to survive, so they are THRIVING. And then your whole lake turns to jelly.

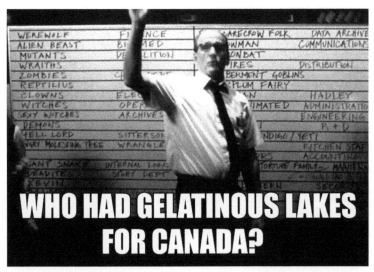

WHO HAD GELATINOUS LAKES FOR CANADA?

As much as we can try to map out the consequence of losing one species (or tens of thousands or MILLIONS of species) we probably won't get it right. Turns out we're really bad at systems thinking. "Unintended Consequences" should be the name of the human sitcom.

One of my favorite anecdotes about systems thinking is when the UK's Royal Air Force airdropped 14,000 cats onto the island nation of Borneo. Why did they do this? Well, it all started when they tried to contain a malaria outbreak back in the 1950s. Malaria is spread by mosquitos, so they sprayed the infamous pesticide DDT (now banned basically everywhere) in an effort to kill the disease-spreading bugs. And it worked! Malaria rates dropped. But suddenly people's thatch roofs started caving in. Well, that's weird. Ahhhh, it's because the DDT also killed a parasitic wasp that kept populations of thatch-eating caterpillars in check. No more wasps, your roof becomes caterpillar food. But that was a minor issue. Because the DDT also killed a bunch of flies, and lizards love eating flies—alive or dead. But the insecticide weakened the lizards, and they became easy prey. And guess who loves to eat enfeebled lizards? Cats. And then . . . surprise! The cats started dying. And you know what happens when your cat population gets decimated?

Rats. Lots and lots of rats. Rats that caused outbreaks of sylvatic plague and typhus. Which sounds even worse than malaria to me. So the World Health Organization said, "Fuck it, let's get these cats suited up."

And that was a cautionary tale about trying to engineer the ecosystem. The point is, we don't know exactly which species collapse will lead to dropping cats on Borneo. But we do know that when everything is interconnected, it doesn't take much to break the whole system. And there is no one to drop a bunch of cats on us when it happens.

Here's another fun one! Have you heard of cannibalistic giants? No, I don't mean something from Scandinavian folklore. Cannibalistic giants actually have to do with marine ecosystems.

There are some cannibalistic fish populations. Silver hake is one. Perch is another. Usually their diet is only about 10 percent cannibalistic, but when their food sources start to dwindle (because humans are overfishing and warming waters, yada yada yada) they all decide eating each other is as good a plan as any. So the fish willing to eat their cousins start doing that, and they start getting bigger. And bigger. And the fish that are too small to eat the other fish get knocked out of the evolutionary game. A population of fish goes from a range of small guys, medium guys, and big guys, to bigger and bigger until we get: cannibalistic giants. These giants then start eating everything, not just other hake. Just a few giant cannibalistic giants can dramatically affect the entire marine system down to the lowest level of phytoplankton.

Okay, sorry, this meme is wayyyy better over here.

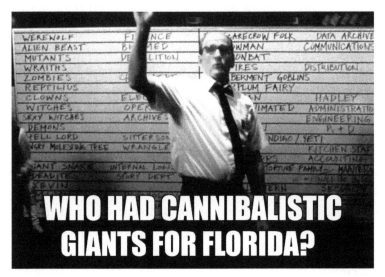

This is called a trophic cascade. You add top-level carnivorous predators to an ecosystem, which causes a decrease in the herbivores beneath them (fish that just eat the plankton and plants), which causes a massive *in*crease in the plants and phytoplankton, which might sound good, but it's not. Because some phytoplankton are toxic algae. And now we have massive toxic algae blooms that end up killing corals and fish and birds and even humans! And guess what else they do? Die and decay and eat up all the oxygen in the water!

I'm not saying cannibalistic giants are going to collapse the entire oceanic ecosystem, it's just another example of how delicate all these things are and how easy it is to throw these systems out of balance.

And due to deoxygenation and dumping fertilizer byproducts into the oceans (and possibly also cannibalistic giants) we're already seeing these massive, toxic algae blooms. There is a name for this too (that's less exciting than the Hannibal Lecters of the ocean): eutrophication. When a body of water gets too many nutrients from all the fertilizers or farming byproducts or sewage we dump in there, phytoplankton grow like gangbusters. There's an algae bloom heading to Florida right now that's five thousand miles wide. That's almost twice the size of the entire United States.

And when these billions of organisms have eaten up all the sewage we dumped in there, they die. And then they are decomposed by other bacteria that eat up all the oxygen that all the other fish we love love to breathe.

And you wanna guess what that gigantic toxic algae bloom smells like? Rotten eggs.

Now let's think even bigger. Have you ever heard of the Amazonian hydrological pump? Me neither. But apparently, all the trees in the Amazon—all 390 billion of them—act like super effective water pumps. They suck up water through their deep roots and release it into the air through their leaves in a process known as transpiration. On a typical day, the Amazon releases 20 billion tons of moisture into the air. That's like 100 gallons per tree! That moisture becomes clouds, which make rain. But this rain doesn't just fall on the Amazon. Wind moves the clouds, and this hydrological cycle provides the water that feeds the crops that feed . . . the entire continent of South America.

According to some experts, deforestation of the Amazon beyond 80 percent remaining of wherever it was before they started cutting it down would likely break this pump. Wanna guess how much is left right now? Eighty-ONE percent. And when that tipping point gets passed, we enter a vicious cycle of desiccation that could turn the Amazon into a woodland savanna. Fewer trees make less rain make fewer trees . . . you get the picture. Already, the dry season is longer, and rainfall has dropped by a quarter in some places. When the rain does come, it's often shorter and more intense, which leads to flooding, which means the land isn't able to absorb the water it desperately needs to continue the cycle, and it's lost to the salty ocean forever.

The real irony is that the farmers who are clearing the Amazon to grow soy or raise cattle desperately need the rain to sustain their farming. But the more they cut down, the less rain they get. AND the less rain the ground can hold because there are less complex root systems, so the water just peaces out. And before you get all ethnocentric again, it's not just South America that relies on this hydrologic pump. These rain patterns also provide up to HALF the water that feeds the western United States. No Amazonian water pump, no American breadbasket. Remember when we were talking about the world running out of water? Yeah, this is that.

And since we're talking about mass extinctions, it's probably relevant to mention that the Amazon is home to 10 percent of all the species in the entire world—including 2.5 million species of insect. Oh, and then there's that whole thing where 390 billion trees suck up a whooooole bunch of carbon from the atmosphere. Without those trees, global warming gets worse . . . no matter where you live! Man, these positive feedback loops just won't quit!

There are a million different things that could happen, a million tipping points that we may or may not hit. Everything may already be fucked, and there's nothing we can do. We could be living in a nearly dead world with no biodiversity plagued by food shortages and water shortages and global famine and global system collapse. We could heat up the world enough so fungus can take over our brains and turn us into flesh-eating zombie monsters. The oceans could become cesspools of toxic gas that kill anyone who comes near them.

We can make up a lot of fun names for these different tipping points like Aquacalypse Now or Insectageddon, or Toxici-tastrophe or Fungistruction or Mushroomalamity? Canfield Calamity?" I dunno. Stagnation station? Dessicadisaster? I give up, you think of one.

But it's going to be a sad, sad world without all the beautiful fish and birds and bees, and even those terrible mosquitos that eat me alive every summer. I love you too, mosquitos. And if we're left with cannibalistic giants in a hot, dry, dark, hungry, swampy, toxic nightmare? It's going to be a whole lot sadder.

But before you think about signing up for Elon's first one-way trip to Mars, let me assure you: Whatever hellscape this planet might turn into, the worst-case nightmare scenario will 100 percent, no questions asked, be more hospitable than fucking MARS. I'll take poisonous Canfield Oceans on a planet with a breathable atmosphere before I move to a dead rock in the vast, freezing vacuum of space that's 34 million miles away. And whatever billions and billions and likely trillions of dollars it will take us to get there would go a whole lot further fixing this incredible planet we've already got.

No matter how this all plays out, one thing is for sure: the path we're on isn't a sustainable one. No matter what the world looks like in fifty years, I can promise you it's not going to look like this. I said it before, but it bears repeating. This could be 5 years, 50 years, or 500 years. Either way, it doesn't really matter. Because if we know it's fucked up and unsustainable, and we know that our kids and their kids are going to directly feel the consequences of this (along with plenty of us alive today), then how can we *not* do something?

Is it worth it to just sit here and wait it out? Order our Uber Eats and keep the thermostat at 68 in the summer and keep partying like it's 1999? Shrug our shoulders and say, "Welp, it wasn't too late in the '70s or the '90s, but it's too late now, so oh well! Let's make some margaritas and watch the world burn!"

I mean, you do you, boo. But shouldn't we at least . . . fucking try?

How did we get here?

Sounds Like an Anthropocene Problem

Hi. It's me. I'm the problem, it's me.

—Taylor Swift

Phew. That section was intense, I know. But whether or not you think it's alarmist or overblown or you don't think any of those things are going to happen in any way that's remotely going to change the way life looks like for the average person here on planet Earth, it's worth investigating exactly how we got to where we are. Just in case. If there's one thing I know about humans, it's that our hindsight is 20/20! Just kidding. I think the quote goes, "Those who cannot remember the past are condemned to repeat it." That sounds more like us.

My husband is a middle school teacher, and he likes to ask his seventh graders, "Could Einstein make a sandwich?"

And of course, they all laugh and say, "Duh, he was Einstein, anyone can make a sandwich." But maybe he couldn't. And neither can you. Because in order to make a classic ham and cheese sandwich with mayo and mustard (and maybe a tomato), you need to know how to grow wheat, tomatoes, and mustard; how to raise cows, and pigs, and chickens; how to milk the cows and turn that milk into cheese; how to slaughter and clean the pig to get the pork and how to cure it to make ham; how to press cooking oil out of some nuts or seeds to make the mayo (along with the eggs you get from the chickens); how to turn mustard seed into mustard; how to make vinegar for the mayo *and* the mustard (which involves also knowing how to make alcohol). Oh, and after your wheat is grown, you need to know how to harvest it, mill it into flour, and bake that flour into bread. And find some yeast out in the wild along the way.

Do you know how to do all those things? Any of those things? Me neither.* Because the division of labor from capitalism, in combination with seemingly endless fossil fuels that allowed us to grow so precipitously and live so far away from where the food was actually produced, means most of us are laughably ignorant about how any of it gets made. The food just appears in the grocery store. And it's always there. All forty thousand products. We can get any fruit or vegetable in any season because we ship it from the other side of the planet that summers during our winters. We can eat watermelon in December if we want. And we don't have to raise a chicken, or kill a chicken, or pull the feathers out of a dead chicken—or pull the innards out of a dead chicken. It's already nicely cut and all wrapped up and ready to be thrown on the grill!

Thank you, Jesús.

Something tells me the vast majority of people who eat meat every single day would give it a second thought if they had to kill and clean the animal before they ate it.† It's like the people who love calamari, but not the ones "with the tentacles" because it reminds them too much it used to be a sea creature. Well, I hate to break it to you, but they all had tentacles. They are all squids.

Here's a Quora question from a real human being to give you an idea of how little people understand where their food comes from these days:

> **Why do some people cruelly kill animals for meat instead of just going to the supermarket to get it?**
>
> 🖉 Answer   ⓐ Follow · 2   →👤 Request   ⓘ   ▢   ⇩   ⋯

I've got some very upsetting news for that lady.

The hard truth is: if something were to happen to the agriculture industry, or to our well-oiled society in general, most of us would be utterly incapable of supporting ourselves. This potential collapse is

---

*   Except the booze. I do know how to make booze. Thank you, South African booze ban.

†   For the record, I love eating all the meats, though I'm working on limiting it to a couple days a week.

something preppers like to call SHTF, or when the "shit hits the fan." And if you recall the short-term piece of the last section, there are lots of different levers that could get us there pretty quickly in our insanely complex yet also insanely fragile system that's propped up by wildly unsustainable levels of global debt based on the future availability of an insanely finite, and nearly depleted, resource.

But there are just a few critical things we need to recognize—that we need to remember—as a species that will help us start to put the wheels in motion to get ourselves back on track.

Our ecosystem has a finite carrying capacity, and that's a fact. And before the dawn of colonialism, it's no secret that Native Americans (and plenty of other indigenous peoples around the world) managed to live in harmony with nature for tens of thousands of years. We had a fundamental tether to nature that every human culture and group understood. We understood the cycles of drought and flood and birth and death; we relied on the planet for everything. We understood that we are not just utterly dependent on the environment—we're part of it. We ARE nature. We are all completely inextricable pieces of a single system where nothing acts in isolation. Every human, animal, and single-celled organism is part of the regenerative capacity of the Earth. The only solution is to keep the system balanced.

It's also no secret that lots of those early humans died when they started living beyond those limits. In Charles Massy's beautifully written book on the future of (and necessity for) regenerative agriculture, *Call of the Reed Warbler*, he notes that "overgrazing, coupled with mismanagement of alluvial river-valley soils, has led to the collapse of virtually all great civilizations of the past—some twenty-six at least."[70] This includes the Mayans and the Egyptians, btdubs. Too many people sucking out too many resources until the land doesn't have enough to give, and a whole bunch of people starve to death before everyone else peaces out or gets enslaved by some other dudes with more food. Tale as old as time. We also have lots of examples of societies who managed to live in balance with the carrying capacity of their land, but most of these were in response to overshoot that wiped out a chunk of their population, took them down a peg or two, and reminded them who's boss (hint: it's not us).

You can roll your eyes and say "living in balance with nature" is all some hippie dippie shit, but it's just facts. Nature isn't some thing you go hiking in on a weekend when you're feeling active or a thing that makes your backyard prettier. It's us. It's the whole fucking thing. And this wacky idea that our ecosystem is a delicate and finite (though renewable) resource with which we are inextricably intertwined as humans is not even a new one within the span of industrial capitalism.

Process philosophy posits that our planet and our food and our economy are fully interrelated processes that cannot be separated. It was started by a dude named Alfred Whitehead back in 1920. It's dumb it even has a name since it's not a "philosophy," it's our state of existence that existed long before Whitehead gave it a moniker. And people haven't stopped reminding us over the last century of this very obvious fact we keep ignoring. The groundbreaking report detailing the carrying capacity of our planet in relation to human development, titled "The Limits to Growth," was published in 1972 and was immediately dismissed as another Malthusian prediction. And the ecological economics movement was in full swing at the time. We were busy ignoring them too.

Process philosophy (and ecological economics as a whole) argues that "there is urgency in coming to see the world as a web of interrelated processes of which we are integral parts, so that all of our choices and actions have consequences for the world around us." Sounds pretty obvious to me (and every single human who lived before the advent of complex civilization), but somehow it's become a radical idea.

We've been living in this disconnected utopia of ease and convenience for so long, we've totally forgotten these very basic truths. Only two percent of us have anything to do with growing the food we eat. We don't have to worry about where the food comes from or know how to make the food because we believe it will always be there. Because it always has been as long as you or I or our parents can remember. The last seventy years have become an unquestionable state of existence for humanity. But I'll remind you once again, we're living in the blink of an eye.

You've probably heard me use the word Anthropocene or anthropogenic a couple of times. What does that mean exactly? Well, the Anthropocene Epoch is an unofficial unit of geological time. It's the period of time we're in right now where humans have completely dominated everything to the point of affecting the planet's climate and ecosystems (and rocks, because that's what geologists care about). While the official geological time period we're in is the Holocene, which started at the end of the last ice age around when we all started farming, the Anthropocene is human-created by definition. The beginning of the Anthropocene is debated, with some putting it at the beginning of the Industrial Revolution, and others putting it around 1950 when population growth and fossil fuel burning went absolutely bonkers. But either way, it's dated to when we started making a visible dent in the ecosystem we're meant to live in balance with.

Unfortunately, one of the defining characteristics of this period is our complete detachment from any understanding of how what we're doing is affecting the planet we depend on. Because if we understood it, I'm pretty sure we wouldn't keep doing it. Or we're just telling ourselves it's fine because it's the only way to keep the Incredible Everything Machine humming.

Massy calls this outlook the 'mechanical mind.' It's a way to talk about our disconnection from the system that we live within. The mechanical mind approaches everything as a problem to be solved with math and machines. It looks at the Earth and the ecosystem as dead ground, a landscape to be beaten into submission without fear of consequence. It traces a line back to the first time we tied up an ox to do our farming and it follows that line straight through to the 70 percent of the Earth's surface we've altered (or razed or accidentally turned into a desert) in our desperate attempts to make more money and food and accommodate ever more humans.

In theory, when the rapid growth of fossil-fuel-powered capitalism started pushing against the carrying capacity of the planet, we would have started pulling back, understanding that what we were doing was worse for us. Worse for our family, worse for our tribe, whatever. Just like killing our babies, or sending grandpa out on an iceberg to die, or shifting from being peaceful societies to warring societies. We would make societal changes to remain within the carrying capacity of the system that supports us. This is the core tenet behind cultural materialism (it's Marvin Harris's whole thing): what are the material changes in our structures and infrastructures that dictate our culture and behavior in society?

But with seemingly unlimited energy and the ability to transfer the costs of growth to ever-more underdeveloped economies, there's been nothing to stop us from growing. We eliminated any of the biophysical restraints that were built into pre-industrial or even early-industrial small-scale communities. We lost the ability to see that we were hitting any limits at all. The energy came from the oil; it made everything possible. And now, even though plenty of activists and NGOs and researchers and climatologists and Swedish teenagers are sounding the alarm bells, we just can't listen. It's the boy who cried peak oil, the boy who cried overpopulation, the boy who cried Plimsoll line again and again.

And I guess I get it. This whole time we've been growing and developing, so many people were getting so much less poor. How could you even be mad? Why would we stop? Obviously, the system works! And there was always more innovation and higher crop yields, and Malthus was wrong, and Ehrlich was wrong, so how could it not be this way? And then we figured out plastics in the 1950s, and industrial food processing, and you could make a cake out of Jell-O and fill it with hot dogs and call it a dessert. It's been nuts!

But to justify this system, we've had to believe in our divine right over every bird and tree and river, over every inch of the planet. We've had to tell ourselves that planetary resources are infinite, that we're not affecting the ecosystem with industrial farming and billions of tons of toxic pollution, that chickens don't feel anything when they spend their entire, miserable lives in a twelve-inch cage, and that we are the masters of all. All this world domination has really gone to our heads.

We've been ignoring the relation of each organism to its environment, and we're ignoring the intrinsic worth of the environment itself. But we thought we could do it because we'd figured it all out. We figured out how to tame the Earth rather than take the time to understand how any of it really worked.

The issue is just that mayyyyyybe we were wrong. And I know nobody likes being wrong. But when you think about how recently we started to engineer the whole thing, it's okay we fucked up. It's okay we were wrong. Good try, everybody. It's only been like a few decades, nothing to be embarrassed about. We didn't exceed the carrying capacity of the planet until the 1970s. Industrial farming (and Jell-O cakes filled with hot dogs) only really got started in the 1950s. We only just discovered our microbiome—the two to six *pounds* of microorganisms that live in our bodies and have the power to affect our mind, our willpower, our decision making—like twenty years ago. How could we know how to make the right food if we still didn't even understand how we eat? All of this science is pretty fucking new. I mean, it would be crazy if we got it right the first time!

sophia
@pastoralcomical · Follow

it's crazy that they only figured out tectonic plates in the 60s. a child in the 50s would say "it seems like south america and africa would fit together" and his mom would go "that's cute honey would you like a cigarette"

10:54 PM · Nov 11, 2021

♥ 182.7K   Reply   Copy link

Read 214 replies

We got super excited when we figured out how to genetically modify crops and how to make fertilizer out of natural gas to make crops grow ten times faster than they used to. It just turns out that wasn't really the best idea. Turns out massive amounts of pesticides kill all the complexity in the soil that is a crucial part of the soil growing things (and storing carbon) in the first place. It turns out that nature is actually really good at being in balance if we just, sort of, get out of the way. The more we try to manage and simplify incredibly complex systems, the more we seem to break them.*

The Anthropocene is defined by man's dominion over nature, but it was never true. Razing the Amazon or Anatolia or the American Midwest was never going to work. Burning fossil fuels to falsely inflate the carrying capacity of the planet was never going to work for long.

For as smart as we think we are, we're actually pretty stupid. What was that whole thing Socrates said? The smartest guy in town is the one who can admit how little he knows?

At the end of the day, we're just another species maximizing resources who happened to stumble on the biggest energy resource of them all . . . and the tide is going to turn. As Harris says:

> The intensification of any given mode of production will eventually reach a point of diminishing returns—i.e., the producers will get less for the same or greater effort. Exceeding the point of diminishing returns inevitably leads to lowered life expectancy and standards of wellbeing and to irreversible resource depletions.

Irreversible. You can't just keep taking more and more and more. No system in all of existence has ever worked that way. It's pretty insane we thought this one could. One day the soil will die, and we'll say we don't know why, and the few humans who are left will probably forget everything we learned about the one thing we're never supposed to forget.

Nature is beautiful and complex and resilient and self-healing, but you can kill it. And countless societies have risen to power and expanded their populations before hitting those limits and falling into decline. And we're next. Understanding those limits and working within them is the only way to continue to survive. Even if you think animals don't have feelings (they do), or they're not as important as human feelings,

---

* Please refer to the skydiving cats of Borneo.

we have to remember—and I can't believe I'm saying this—they're still valuable as more than just food. As theologian and environmental ethicist John Cobb says, "Human beings may be the apex of the biotic pyramid from which they evolved, but an apex with no pyramid underneath is a dimensionless point." Living at the top of the food chain doesn't really matter if there isn't anything left at the bottom.

And if we don't stop to understand how we've gotten so far from where we started, we may be destined to watch it all unravel.

Sounds Like the Oxen's Problem

**Be fruitful and multiply, and fill the earth and
subdue it; and have dominion over the fish of the
sea and over the birds of the air and over every
living thing that moves upon the earth.**

—Genesis 1:28
The effing Bible

The story of human civilization is the story of surplus. As soon as we figured out how to store that very first wheat surplus, it instantly became about more. And once we could *more*, we started *more*-ing as much as we could. We're still more-ing as much as we can. Not because we're greedy or unique. Every species on Earth will find as much energy as they can so they can have as many babies as possible. It's just a biological imperative.

But boy, was it slow going at first. We worked for thousands of years to get better at farming the types of food we could store. We took branches and sticks to mix up the soil to help more stuff grow. We were learning. But there was one invention that changed the name of the game for the rest of time: the plow. We had finally figured out how to really jumpstart that whole "getting more" thing . . . but there was just one problem. It meant hooking up this new-fangled plow to an ox. And unfortunately, we still worshipped them.

Before the plow, and looooong before that long-haired hippie in Birkenstocks started turning water into wine, most cultures were animistic. We all believed in that spiritual Gaia mumbo-jumbo about the sacredness of the planet and all her beings. Rain gods and tree gods and oxen gods, yada yada yada. But social philosopher Daniel Schmachtenberger argues that our belief in animism and worship of nature died with the plow.

Because when the plow was invented, it meant using an ox all day to pull it. And if you worship an ox, it isn't really cool to turn him into your slave. It's the same reason why human slave owners had to believe that Africans weren't really humans—in order to justify subjugating them. Man's dominion-over-nature ideologies had to follow the invention of the plow so that men could justify beating up animals all day long.

Any peoples who wanted to grow and have more food had to shift to this ideology. But if any culture said, "Nah, we really love these oxen, and we don't want to hurt them," that culture would have been wiped out by the now dominant cultures with all their extra food and larger populations. Adopting irrigation-grain agriculture (and the plow) meant you suddenly had enough calories to feed an army of several thousand men. And the oxen-lovers did not. And the bigger the surplus they accumulated, the bigger their societies could grow. More food, more babies, bigger armies, more power, more wealth. Tale as old as time. This kind of technology is considered obligate. As in, a culture is obligated to adopt it, or they will cease to exist.

What other tech has been considered obligate since the plow was invented six thousand years ago? Well, all kinds (the gun comes to mind), but I would argue it's both technologies and ideologies that have been obligate. If you believed nature was sacred and wouldn't touch it or harm it or raze it, then your country or tribe was not going to make it to the top. If you weren't willing to kill other people in a war (or you didn't have a gun to do it), then your people would be killed. Essentially, we may have just killed off all the peaceful people who were dumb enough to let us come in with our guns and our smallpox.

It's like the dodo bird. The Darwinian story I grew up hearing was that the dodo bird was "so dumb it went extinct." But that's not really the whole story. The dodo bird was so friendly it was always walking right up to men with guns who shot them until they went extinct. They didn't exactly get "natural selectioned" out. So much for being outgoing.

There are some other, more complex factors at play here, of course. Some cultures would have gladly given up their animism to adopt the plow, they just didn't have the right animals to do it. Most of the large animals in West Africa, for instance, can't really be domesticated or aren't big enough to plow the fields. These cultures continued tilling their lands by hand with a hoe, and they continued living in harmony with the bush on a relatively restricted energy base compared with all those plow-happy societies. Other cultures may have had the right animals, but the crops they could grow just didn't store well, like yams and bananas in South America, which means no surpluses.

According to Harris, this explains why societies who had been existing in Africa and South America for tens of thousands of years were "weaker, less centralized, and more egalitarian than their European counterparts, and why it was ultimately the Europeans who developed capitalism and enslaved the Africans rather than vice versa."

So now we're left with a planet largely devoid of cultures that exist in harmony with nature because the ones willing to destroy it—along with any other humans in their path—are just going to keep winning. And eventually we see colonialism spread all over the globe wiping out perfectly peaceful people and replacing them with musket-wielding capitalists.

However, contrary to the existing narrative (written by the victors, of course), plenty of societies decimated by surplus-happy invaders were by no means backward or savage or stupid. More and more evidence is coming to light that ancient cultures even before the dawn of widespread agriculture practiced a mix of farming and semi-no-madic hunter-gathering, had complex social and political structures specifically designed to avoid little despots, built massive monuments, had wild fucking parties, and maybe had found a pretty rad way to enjoy life and live in harmony with nature (and human nature) without creating nation-states endlessly vulnerable to bad actors, power grabs, and ever-worsening inequality. They chose the structures of their societies—they weren't just too dumb to invent capitalism. And this was happening well before the accepted narrative of "seven thousand years ago we got smarter and made cities."

So, we lost the plot from being big, blue, eight-foot-tall, nature-loving creatures swinging from the trees of Pandora and turned into evil capitalists like Giovanni Ribisi raping the planet to mine for unobtanium. Doesn't sound great. But it's worth digging a little deeper into that last jump from "humans created writing and complex civilizations" to "eight corporations own everything on Earth" in order to understand where exactly we went off the rails—and how we might get back on track.

To do this, we have to follow the trail of the surplus.

All economic philosophies answer the same question: who owns the means of production? Who owns the land that makes the food (or the factory that makes the money to buy the food)? In capitalism, it's privately owned; in feudalism, feudal lords were each granted their little piece. In socialism, it's communally owned; in communism, it's owned by the state. But in plenty of those early, semi-nomadic, semi-agrarian cultures, nobody really owned anything. Certainly nobody owned THE LAND. How could you own land? Or a tree? What a weird idea. But once we started farming in the same place all year, it was like, well, this tree wasn't just here. I put it here. So now it really is my tree. As soon as we got that ownership, it meant private property, which meant wealth. As soon as we started using oxen and plows, we got even more wealth. And as soon as we started stockpiling that wealth, we got ruling classes and the inequality that comes with them.

Inequality in any society can be thought of as the percentage of surplus created that goes to the few instead of the many. The only way to accumulate wealth is to control the means of production that creates the surplus, and then keep that surplus for yourself.

The first Sumerian priest-kings (c. 5500 BCE) claimed to talk directly to the gods to bring a more bountiful harvest, so obviously we gotta protect them and worship them and give them all the nicest stuff.

In the Indian caste system (started around 1500 BCE) you were good in a past life, so you got to be at the top, and the people at the bottom of the system were bad in a past life, so they had to do hard labor. Of course, this was just invented by the Aryans (OG Persians) who invaded India, put themselves at the top of the system, and pretty much made everyone else their slaves. Tale as old as time.

For these early societies to be successful, they also had to exist in a place where the peasants couldn't easily escape into nearby areas and survive on their own without experiencing a significantly worse quality of life. So long as I can still grow a little food and support my family, I'd rather be free than live as your subject. But the more food and land was controlled, the harder it was to survive on your own, the more power these leaders commanded and the more surplus they could keep for themselves. Six thousand years ago, you did what your leader said because he was the only one who could talk to the harvest gods. Now you do what your boss says because he is the only one who can sign your paycheck. Without the army of the ruling class controlling the grain production, or the means of coercion through implied starvation when you lose your job, it all falls apart.

However, while most human civilizations have been based on a working class doing all the work and a ruling class controlling who gets how much of the surplus, both power and wealth distribution have differed wildly across cultures over time. The narrative that we as humans are destined to serve our leader's bottomless pit of greed and over-accumulation is a recent one.

When you think of the Aztecs (c. 1300 CE), you probably think of gold-drenched kings and pyramids (and human sacrifice). But in the Aztec city of Teotihuacan, they tore down their pyramids and replaced them with equitable public housing for all. For real. Instead of despots, many Aztec cities were ruled through complex communal political structures. Turns out, they were actually pretty good at making sure all of their people had food and decent houses and that the surplus was shared more equally among everyone (human sacrifice aside).

Ironically, the Harappa civilization that preceded the Indian Caste system in the Indus Valley (c. 3300–1300 BCE) is thought to be one of the most egalitarian civilizations we've found. Evidence is scant, but most people had the same-sized houses with the same-sized bricks, wore the same jewelry, used the same pots, and were buried with the same stuff. They lived in such an abundant place that they didn't make weapons and never went to war. (Until climate change wiped them out.)

Hammurabi's Babylon (c. 1750 BCE) wasn't egalitarian by design, but inequality was kept in check by periodically canceling everyone's debts. This wasn't an altruistic act; it was a method of keeping society running smoothly. Because extreme inequality causes unrest. Debt, by definition, accumulates wealth in the hands of the elite, and creditors always tended to be greedy. Cancel debt, rebalance the scales, problem solved.

In ancient Pompeii—one of the most unequal cities we know about in ancient history—the upper class was so rich they would spend more on a single banquet than a senior bureaucrat would earn in a year (sound familiar?). But they placated the poorest of the poor with subsidized prostitutes, gladiator fights, and free wine.

Ancient Egypt, of course, is remembered for the pharaohs who made all their clothes out of solid gold and enslaved basically everyone else to build those damn pyramids. But they did have gender equality, so that's pretty cool![†] They also canceled debts, but it was mostly because too many people had become debt slaves, and they needed more workers and soldiers to keep building all that shit.

Throughout history, inequality did tend to widen as societies adopted draft animals and plows, but the smarter rulers found ways to rebalance it. Because the ones that didn't (or distract the peasants with hookers

---

† Fun fact: we're actually closer in time to Cleopatra's reign than she was to the building of the pyramids. The pyramids were completed in 2490 BCE and Cleopatra died in 30 BCE. Which means the pyramids were finished 2,460 years before her death, and the date today is just 2,053 years later. Weird, right?

and wine), would face revolt. And this happened *a lot*. The history of peasant revolts spans more than two thousand years around the entire globe, from Europe to Africa to Russia to China to Japan.

And that brings me to feudalism. After Ancient Rome fell (less inequality, more democracy, zero women's rights, but sooo many slaves), the decentralized power vacuum of the Western world was filled by feudalism. In feudalism, basically everything is owned by knights and lords and barons (granted by the king in exchange for protecting his land), and the other 85 percent of people have to work that land for the privilege of existing on it. Instead of using slaves for everything, now we had serfs! But, you know, serfs were still pretty much slaves. The only difference was that they couldn't be bought or sold, they were allowed to own things, and they were usually given some land to cultivate for themselves.

But if the lord's manor got transferred to a new lord, they came with the land and had to continue to cultivate it for their new boss. They were just sort of there. Like buying a house that comes furnished. And the serfs could scream, *This isn't fair! Why are we sentenced to a life of subjugation!?* But there wasn't anything they could do about it because they still had to eat. So, you know, shut up and get back to work.

While I'm sure that sounds pretty unequal, it's worth noting that inequality was actually quite a bit lower in feudal medieval Europe than it is today. In the year 1340, the top 10 percent of ultra-rich elites owned about 66 percent of all the wealth.[71] Then the Plague caused a massive redistribution in wealth with their share dropping to a low of about 49 percent (because they ran out of peasants to exploit, so the peasants could demand better terms for their servitude). But today, the top 10 percent of humans own 85 percent of all the wealth in the world! In 2017, the richest eight individual people (*cough* white dudes) owned as much as the bottom 50 percent of the planet combined.[72] Something tells me that number has only gotten worse since then. Especially when you consider that from 2020 to 2023, the richest one percent grabbed nearly two-thirds of all new wealth created—$26 trillion for them and just $16 trillion for the rest of us.[73] We're actually living in the most unequal society in all of recorded history! And apparently feudal serfs had wayyyy more vacation time than the average American![74] Fun, right?

Anyhoo, other than all that human subjugation and rampant inequality and all those religious crusade murders, medieval feudalism seemed like a great system for like six hundred years. Power was more broadly distributed in smaller kingdoms, the lords and knights and landed gentry got to feel powerful, the serfs worked in exchange for food and protection (and definitely weren't slaves), the king got to keep his kingdom, and everybody got to eat. Who knows, if the Black Plague hadn't killed half of Europe, we may still be living in it!

Just kidding. Turns out, the serfs were pretty unhappy with their whole setup and had already been revolting (mostly unsuccessfully) for hundreds of years. But in the wake of the Black Death (and a bunch more rebellions), they finally got what they wanted. Feudalism was largely dismantled, and all the serfs got to work land they owned in a subsistence lifestyle based on sharing common resources with the rest of the villagers, like land for grazing cattle and forests full of rabbits and trees. Wages went up—way up—and everybody took five months of vacation a year because everyone was happily working the bare minimum they needed to live! Hooray!

Except, the people who were used to accumulating all the wealth from other people's slave labor weren't super pumped about this. So, as a reaction to their 'chronic disaccumulation,' they sort of accidentally invented capitalism. On purpose. To take all their land back.* The "commons" that were shared by all the peasant villages were enclosed and privatized; the peasants were evicted and then murdered if they had something to say about it. As we know, capitalism requires cheap labor to work, and now suddenly they had lots of it. All the peasants were starving since the people in charge took their only source of food, so now they had to work for whatever wages the new owners of the means of production were willing to pay. Which—shocker—was not a lot (sound familiar?).

And all the while, the people in charge weren't just "enclosing" land and making wage slaves in Europe. They were also colonizing the rest of the world, making millions more actual slaves, extracting and accumulating *trillions* of dollars of natural resources, AND going to war with each other. Which means they had to keep inventing better weapons to be sure some other guy wasn't gonna steal all the land they just stole. Man, they sure were busy. The decentralized powers of feudalism thus transitioned into the consolidation of power in absolute monarchies and the overwhelming concentration of wealth to the creators (and perpetrators) of capitalism and colonialism.

Long story short, the plow was invented so humans could grow more food with less work, and animism died out as a religion and was replaced by religions that saw man as above nature as a way to justify the subjugation of animals for human benefit. The cultures that adopted the new religions allowing them to store the most surplus energy were the winners. The surplus led to division of labor so more people could invent things, which led to technological advancements like better weapons. The surplus also led to private property, which led to wealth, which led to ruling classes, bureaucracies, nation-states, then empires, then colonialism (and now late-stage capitalism)—with ever-worsening

---

* Check out the opening chapter of Jason Hickel's *Less Is More* for a more detailed ride through peasant revolt, success, and re-disenfranchisement.

inequality as the economic elite extracted "free" value from the land and human labor in a never-ending game of accumulation. Any societies that still existed in harmony with nature (or who didn't have guns) were subjugated, eradicated, and called savages. Any socialist peasants left fighting against the system were systematically murdered and called Godless vagabonds. And anyone left over was forced to labor for the capitalists since they didn't have the guns to beat 'em.

That's about as short a paragraph as I can write about that.

So our new cultural imperative was believing in our God-given right to take whatever the fuck we want, accumulating as much shit as we could, and murdering anyone who got in our way. Cool.

Sounds Like a Personal Problem

**I have a joke about trickle-down economics.
But 99 percent of you will never get it.**

So now it's like 1750, and white Europeans have pretty much taken over everything, enslaved a whole bunch of people, and capitalism is the bee's knees. We've changed the mode of production from hunting and gathering, to subsistence farming, to slave-powered city-states, to feudalism, to agrarian capitalism, to industrial capitalism (laboring in a factory to buy the food that someone else farms).

The serfs are finally free from the bonds of their lordships and allowed to make their own way in the world. Now they get to be poor and miserable, but it's their own fault! At least they could blame the station they were born into for their inescapable poverty. What's your excuse? And the people who had already been owning the stuff for centuries were the ones making the rules about who could own stuff and how, which gave them a significant advantage in the race to own the most stuff!

But while basically all of our previous complex societal structures going back a couple thousand years were based on ruling classes creating varying levels of inequality, and the version of capitalism we're in now is the most unequal society in all of human history, it may surprise you to learn that capitalism was actually supposed to be . . . different.

And once again, we sort of have Adam Smith to blame for what happens next.

Smith may have accidentally penned the theory that negated the entire value of energy and nature in our economy (oops), but the beginnings of capitalism that Smith sort of invented and so fervently defended were meant to improve the lot of the poor. I mean, not as well as that whole post-feudal proletariat paradise where we all chose to share common land and take five months of vacation a year . . . but once capitalism was entrenched, and all the socialist peasants were beaten into wage-slave submission, capitalism did sort of work for a minute there. In the days of mercantilism, it was seen as a way to make everyone richer. It was better for everybody, not just those eight white dudes.[†] And it was certainly a whole lot better than being a slave peasant.

Obviously, capitalism gave us all kinds of progress and innovation that wouldn't have happened otherwise. (Why would a serf waste time inventing something?) It brought us out of the disease-ridden stench of shit-filled streets into this miraculous era of ease and convenience (and toilets and bathing). And the more people who spent money, the more money there was to go around, the more poor people had more money.

It was the only time trickle-down economics was actually a thing (sort of). Rich people spent money on stuff that enriched the state that meant the state could provide more services. International trade opened up markets making more things available that weren't available before, which meant more people could make more money. And at this time of extreme national isolationism, that was huge. The more people who had money, the better off everyone else was. More shoes! More shirts! More service![‡]

[†] I'm skipping over a sizable chunk of early post-feudal capitalism where everyone was intentionally impoverished for hundreds of years, living standards plummeted, and soooo many people died. But after that part.

[‡] Coincidentally, lending money to the government to provide public services (and go to war) was a way for aristocrats to entrench power in shifting societal systems. Until they kind of ran out of money, and the newly wealthy merchant and banker classes stepped in, creating the powerful "merchant aristocracy," a precursor to the crony capitalism and corporatocracy we see in capitalism today.

This was all part of an oft-repeated goal from Smith's economic theory known as "universal opulence." That might sound like everyone should be living in mansions and swimming in diamonds, but the goal of universal opulence wasn't based on the inherent value of owning stuff; it was part of Smith's larger vision for a more civilized society. Universal opulence means that rising material wealth was better for the social and moral fabric of society. The core tenet behind capitalism is rational actors acting in their own self-interest. But as strange as it may sound, Smith's original vision of capitalism was *only* about using that self-interest as a lever for social good. He wasn't an economist; he was a sociologist.

The inherent desire to accumulate wealth helped every person develop beneficial virtues like self-control, the ability to delay gratification, and considering the needs of others before you act. Smith believed that the market could be structured so that the pursuit of self-interest that drove our endless desire for surplus and "universal opulence" actually made communities better and stronger. The obsession with luxury and buying ever-finer goods was just an unfortunate side effect.

The only way to make money to buy the things you want is to participate in your local market, selling your wares to other people in your community. But participating in that community market worked to make everyone a better person because—and this is important—if you acted a fool, nobody would buy your shit anymore. Your self-interest could only go as far as was still accepted by the community. People don't do things that hurt the larger group because they both need and want to be a part of that group.

I spend my time running my shoe-making shop, cobbling away, not because I'm a philanthropist, but to make money so I can eat. But that work I do puts shoes on people's feet who don't know how to make shoes, and everybody benefits. But running my business means I have to remain a productive member of society. I have to interact with the butcher and the baker and the candlestick maker every day. I can't lie, cheat, or steal, or people will stop coming to my store, and it's worse for me. I don't stay honest because I care about other people, but because I care about myself above all, and I want everybody to like me.

In Smith's model, our innate desire for approval from other people makes us act selflessly. (I still believe people act selflessly for better reasons than this, but whatever.) It was self-interest that happened to work for social good at the same time. When it stops doing that, Smith believed it was time to stop using it, or to change the way we've chosen to structure the markets we've made.

The problem is that the system Smith envisioned doesn't quite work in the societal structure we ended up creating. The community that existed back then to rein us all in has vanished. We're isolated from one another. I can run a business on the internet and never see another human. I don't go to the bookstore; I buy my books on Amazon. I don't know my neighbors. And even if I do, they don't sign my paycheck. I can tell them to get bent with little to no consequence. You can tell strangers on the internet to go eat a dick with absolutely zero consequence. If you don't want to be a part of the social fabric these days, you don't gotta.

Laissez-faire, mon cul.

And just as with the individual, Smith didn't believe corporations would make decisions that were beneficial to society out of the goodness of their hearts either. His concept of public good didn't just extend to the butcher and baker and candlestick maker. He knew that public institutions and businesses alike needed to be kept in check so that they were also always working for the betterment of everyone. He knew if they weren't regulated, they definitely wouldn't. Because, you know, that whole "self-interest" imperative.

Because of this, he believed one of the sole purposes of lawmakers was to make laws making sure that corporations and those wealthy merchant bankers weren't doing things at the expense of social good. And the sole purpose of economic policy was to keep the purchasing power of the economic majority high—by keeping wages high and prices low. Obviously, our GDP-loving economists today have strayed a bit from that ethos. Smith objected to excessive government interference in economic and social life because, during the age of mercantilism, laws were very frequently made to protect private interests at the expense of public interests.

It's kind of hilarious that his vision of a "free market" was borne from the fact that regulations back then tended to favor big business and make it harder for the small guy to compete. Whereas in our markets today, any regulations tend to be ones that restrict what businesses can do, usually because it's better for people. Like regulations against dumping toxic waste in a river or regulations against hiring six-year-olds to work in your factory. But somehow that's been conflated to mean he thought there should be no government interference in business. People started calling him the father of laissez-faire capitalism—a title I'm sure he would fervently disagree with in today's world.

Laissez faire is French for "let it be." It means "hands off." Let the market do what it's going to do because markets are inherently efficient. And any rules and regulations that humans impose on them just make them less efficient. At least that's the argument they're making. But the thing about the "free market" is—it doesn't exist. Markets are man-made. We invented them. No market exists until you make rules about what exactly makes a market. What can people own? America decided a person can own trees and oil and land. Norway decided those things pretty much belong to everybody. We, as a society, decided a person can't own another person (though that one took us a while). The United States decided people can't sell sex or kidneys or cocaine, while some other countries decided selling sex is fine, but you can't sell your womb as a surrogate mother. Markets have to decide on all kinds of rules in order to exist. They need rules on what constitutes a monopoly and when we need to break them up.* Markets need rules on what contracts are, how they're enforced, and who loses out when someone goes bankrupt and can't pay up.

People who tell you free markets need to be unregulated are insane because there is no such thing as a free market. We have to make these

---

*   Monopolies break the efficiency of capitalism, so these must be broken up if you want the market to function at all how it's supposed to, which is a strong argument for why capitalism is an inherently flawed system.

rules, and they have no bearing whatsoever on the "size" or intrusiveness of government. They don't affect how much the government taxes people or how much it spends. The market doesn't function—it doesn't exist—without these rules in place.

Where Smith's vision really went off track was . . . probably around the same time the rest of this crazy shit started happening. Most notably in the last fifty years. While we've always had wealth inequality, as it's inherent to a capitalist system (and apparently every other system we've ever created), we used to have better rules to protect more people. Rules that reined in those tendencies of the system to accumulate power and money to the very few. Hammurabi canceled debt. The mid-20th century had unions and safety regulations and strict anti-trust laws. Smith would have loved all those. He always advocated for better rules that benefited more people. He didn't give a fuck about the free market. He gave a fuck about building a market that worked in the interest of the public good.

In Smith's day, less regulation was better for more people. Today, the opposite is true. Society changes, technology changes, and governments and politicians and judges have to decide how the rules should change to keep up. We didn't have laws about who owns software code before someone invented software. And—this is the important part—over the last half century in particular, basically all of these rules have been made or changed by the owners of the means of production in favor of themselves to the detriment of the greater good. That's all. That's the crux of the problem. It's always been the crux of the problem, and different societies over time have come up with different ways to manage it. And the ones that didn't have gotten French-Revolution-guillotined like Louis XVI.

Inequality just sort of tends to get worse over time without the right controls in place so that all the wealth doesn't "trickle up" to those people making the rules. It's just the way capitalism works. But it's another positive feedback loop. The more money people accumulate, the more power they get, which gives them the ability to make the rules to give themselves more money, which gives them even more power to further change the rules in their favor. This is what we mean when we say "late-stage capitalism." It's the nature of the system when it's left unchecked.

Way back in 1873, Edward G. Ryan, the chief justice of Wisconsin's Supreme Court warned:

> The question will arise, and arise in your day, though perhaps not fully in mine, 'Which shall rule—wealth or man; which shall lead—money or intellect; who shall fill public stations—educated and patriotic free men, or the feudal serfs of corporate capital?'[75]

I think we all know how that one ended up playing out.

Ironically, if this keeps up, the rich dudes are going to break capitalism because a healthy, wealthy middle class is what keeps them rich in the first place. If we don't have money, we can't buy the things their companies make, and the whole thing falls apart.[†] Even though they will be worse off in the long run, they're going to keep accumulating and amassing as much money and power as they can. But I guess they're just too busy Scrooge-McDuck-diving into their silos full of hundred-dollar bills to realize that.

And as if it wasn't hard enough to rein in corporate interests in 1776 or 1873 or 1929, now we have the miracle of globalization to contend with. We can't even make rules to protect people and benefit society in our own countries, and corporations now operate across dozens of countries where we don't make any of the rules. And they just keep growing and growing and growing. And because those businesses have grown so gigantic, so far beyond the communities they used to exist within and depend upon, the whole social piece of Smith's vision falls apart as well. Just like you don't have to concern yourself with how your business affects your neighbor, neither do the eight corporations that own everything.

We went from using the division of labor of capitalism to increase the wealth and wellbeing of so many different people through small community-based farms and businesses to multinational corporations

---

[†] What's that? Congress is trying to push through another tax cut for billionaires in the same bill—in the same breath—as they cut Medicare and social security? Goddamn, they're just getting so ballsy!

completely detached from the locus of the communities they're meant to serve gobbling up profits at the expense of literally everyone and everything. Our rural economies have collapsed, communities have collapsed, and everybody's moved to the city so they can struggle every day to get by while feeling miserable and making heaps of money for someone else—and destroying the planet while we're at it.

And the rampant inequality baked into the system we've got means growing economic insecurity for most people who are working their asses off and barely hanging on. And the more desperate people are to survive, the less trusting they are, the more selfish they become. And the less people trust one another, the more the whole thing falls apart.

Smith probably would have been pretty bummed if he realized that we created a system where all of the things he believed were there to keep people and businesses in check and make us better have been dismantled. And the book he wrote on the topic that has become the most famous book in the history of capitalism is being used as an argument to do exactly the opposite of what it was supposed to do. Woof. Sorry, buddy.

The Tragedy of the Commons

But this complete disconnection from community and the means of production has much more far-reaching effects than trolls on Twitter or cursing out your neighbor.

Imagine there is a tree in your yard. Well, first imagine you even have a yard or own anything at all. If you cut down your tree, you can sell that tree as lumber and make money. And while that tree is no longer working to convert carbon dioxide into oxygen and make our air more breathable, it's not really noticeable. It's just one tree. Now you have one less tree and a little more money, so obviously you are going to keep doing that.

The problem comes when everybody cuts down their trees. Because then the effect is noticeable. Then the Earth warms faster because there are fewer trees helping to suck up all the carbon, and bird populations die out 'cause they don't have any trees to live in. But right now, it's only upside for me to cut down that tree.

This is known as the Tragedy of the Commons. We act in our own self-interest, which means I'm gonna cut down that tree because it's mine, and I can make money if I do. But as soon as everybody does that same thing, we're screwed.

Think about it like the Golden Rule (do unto others as you would have them do unto you). Except in this version, imagine if everything you did, everyone else on the planet did too. What would that planet look like? So you throw some trash out your window, no big deal. If everybody did that every time? We'd be swimming in it.[‡] So you pick a flower from someone's garden, no big deal. If everybody did that? There wouldn't be any gardens. This is known at Kant's categorical imperative, and we can apply it to a whole bunch of things in life. Actually, you can apply it to everything in life. It's the Golden Rule on a global scale.

You should always act in a way that you would want other people to act in society. Do unto others as you would have them do unto you. But also, you should always act in a way that could be generalizable to all people. Before you do anything, ask yourself what the world would look like if everyone was doing that same thing you're doing.

That imperative worked in small societies. You couldn't do unto others in a negative way because you would feel the effects of it when people shunned you or stopped doing business with you. You couldn't cut down all the trees in the forest at the edge of town because your neighbors would be like "What the hell, dude? We were enjoying those trees!"

Unfortunately for Adam Smith, and for Kant and his imperative, everything broke down as soon as markets and businesses became detached from the communities they serve. It isn't just the disintegration of the social fabric that incentivized positive individual behavior. We now have a system that is so large and global and individualistically oriented and simultaneously isolated that the mega-corporations running things have absolutely no need to concern themselves with the communities they're affecting. And once we removed ourselves from the community impact of our decisions, we were all able to cut down the trees without seeing the damage it was doing. Because it was always somebody else's trees.

---

‡　I mean, we kind of are swimming in it.

The Tragedy of the Commons, of course, has very little to do with the trees in your backyard (unless you're a soy farmer in the Amazon). That backyard is our planet, and that person has become corporations and governments. Ironically, the Tragedy of the Commons was originally meant as a critique of socialism (and a justification of the enclosure movement): no person will care as much about a shared resource as they will about a resource they own. And that's exactly how it's working. Because these corporations can cut down as many trees as they want, and they don't have to pay the price for it. It's socialism for big business. Because the pollution from burning fossil fuels or the cost of cutting down trees isn't part of our economic models. Because oil companies don't pay anything for the oil they take out of the ground—that is arguably owned by every member of a society whose land contains the oil.[§]

We've socialized all the risk and negative consequences to be carried by the people who have to live on the planet, and privatized and commoditized the gain to be collected by the CEOs and world leaders and whatever other rich and powerful people control the corporations and the government. It's the Sumerian priest-kings claiming the right to everything because they're the only ones talking to the harvest gods. Except we know the harvest gods aren't real, and no one stopped to ask why we keep worshipping these dudes.

When you think about it, it's our trees and our oil and our soil and our water and our rare earth metals and our environment, but they stole it from our ancestors, called it private property, and we were all forced at gunpoint to agree that nature belongs to whoever sticks their flag in the ground first. The oil belongs to the person with enough money to drill for it. They cut down all the trees, and we get Canfield Oceans and cannibalistic giants. I mean, so do they, but I guess they already have underground bunkers ready for when the SHTF.

The reason we're so blissfully unaware and we can just keep on living without worrying about any of these things that should be incredibly worrying is because—up until now—we could externalize the price

---

[§]   Or, in an even wilder take, is owned collectively by every single human on the planet. But we don't have to go that far here.

we're paying. We can pump our fields full of fertilizer to hide the fact that they can't produce all the food we need. We can take resources from a place we've never even seen and ship 'em across the world for pennies. We can exploit labor and technology to make it seem like everything is just fine so long as there are new places we can exploit (most notably, poor countries in the Global South who are just trying to play catch-up in our game of world domination). And because the world is now so much bigger than our forest or farm or community, we don't see it.

And because the whole system is incentivized toward endless wealth creation, that's the only thing that drives decisions. You try to maximize your wealth by getting a better job, politicians try to maximize theirs by getting bigger donations and staying in power, and corporations . . . well, we know what corporations do. The politicians are going to keep taking big donations from the corporations, and the corporations are going to keep polluting and destroying the planet so long as they can get away with it—and so long as maximizing wealth is the #1 goal of society on both the individual and national levels.

The Race to the Bottom

Unfortunately, it gets worse. Since the guys who "own" the oil and the trees don't have to account for the cost of nature in the economic models we invented, destroying it is the best way to keep keeping more surplus for themselves. Because not only is the environment free, extracting resources from it gives us an advantage over anyone who's not extracting resources. And we never feel the effects of the damage we're doing! Drill, baby, drill! The capitalist utopia we've built not only exploits the vast majority of people working in it for the profit of a few, the systems we've created actively incentivize everyone to extract as much value from the environment as possible. And none of us get to share in the surplus because we didn't put our little flag in the ground first.

And since there's no one to stop anyone from doing that in a dishonest way, or a way that harms the planet or the future, there's no reason not to.

That's what's so crazy about these farmers in the Amazon. They're literally burning down their future to make a buck today.

The thing is, for the longest time, the Tragedy of the Commons just . . . didn't feel that relevant. We were all getting richer; there were enough trees to cut down for everyone. So we went from the wooden plow to the mechanical plow to the Industrial Revolution to the musket murder and colonization and industrial agriculture; it was fine. Because we still weren't anywhere near hitting the limits of the system that we existed within. We could take all the oil out of the ground, burn the coal, kill the buffalo, nuke the Japanese, exterminate the dodo, cut down the trees. It didn't matter. The reason energy and the environment weren't inputs into the early economic models is because they literally didn't affect the outcomes. Human labor mattered. Capital mattered.

For all the plundering we were doing, we weren't making a very big dent. Our blue economy square was still so much smaller than our big, green circle of an ecosystem. The energy and materials were always going to be there. They were always going to be infinite. And the obligate philosophy of "God's infinite bounty to man" led us to believe we were blessed with these gifts for eternity, and they were ours with which to do as we pleased.

And the real rub is, much like the obligate technology of the plow, absolutely destroying nature has become obligate behavior. It's not just individual profit maximization. Nations and corporations alike do not have the luxury of respecting nature. If you don't kill the buffalo or burn the forest, someone else will, and their business or society will outcompete yours. So there's really no reason *not* to do it.

This is what we call a fucked-up race to the bottom.

We could make all these laws for America going green and getting off oil and shifting away from endless growth capitalism—but then it's just more for Russia and China and Brazil and every country who definitely doesn't give a shit about the environment in what they most likely see as a zero-sum game toward world domination. It's a multi-polar trap.

Think about going to a concert. One person at the concert stands up to see the stage better. But then everyone behind that person has to stand up as well, or they won't be able to see. And now everyone is standing, but no one can see any better than when they were all sitting.

And the "trap" part is that now everyone has to keep standing, or they won't be able to see the stage at all.

So if any country right now decides to stop cutting down trees and using oil and building nuclear weapons—all the things that give us money and make us powerful—they would lose (according to the arbitrary objective we're optimizing for). They'd be the only country sitting down at the concert. Now, if everyone stopped doing the thing that was bad for the planet at the exact same time, it'd be fine. We could all get rid of our nuclear weapons on the count of three. But that's not really how geopolitics work. That's certainly not the direction geopolitics are heading as I'm writing this. And even if China or Russia tells us they're going to stop doing something . . . are you going to believe them?

So somehow, everyone in the world has to agree to sit down at the same fucking time.

Sounds Like a Tomorrow Problem

Over the bleached bones and jumbled residues of numerous civilizations are written the pathetic words "Too Late."

—MLK Jr.

At this point, you might be wondering, "But how did we let it get so bad?" If we know we're all on a path to destruction, and we're making a big fucking dent, and everything is unsustainable, and people have been talking about this for so long, how come we haven't done anything about it? Another excellent question.

But it's much more likely you've been thinking, "Meh, we'll figure it out. I don't think it's gonna be that bad. Fifty years is really far away!" Lolz. Not really. The thing is, our society has evolved far more quickly than we have been able to as a species. We didn't start farming and settling in one place until about twelve thousand years ago. Before that, the biggest threats to human existence were immediate ones—like starving to death or a tiger eating you. So the humans with the best near-term focus would survive over the ones lackadaisically strolling through the woods and daydreaming about the future without paying attention to the hungry tiger. Except the tiger we're facing now only *seems* like it's really far away. And because of this evolutionary adaptation, we just don't place that much importance on things in the future. And the further in the future it is, the less we care about it. This is known as "future discounting." It's the nature of our myopia as a species.

Future discounting means that given a choice, most humans would take $100 now rather than $110 in a month. The higher the discount rate for the person, the more money they have to be offered in the future in order to change their mind. How much would it take you to switch from money right now to money next month? $130? $150? And the same is true for harm. Being kicked out of your apartment tomorrow is far worse and far more urgent than being kicked out of your apartment next month. Your brain simply tells you: I don't have to worry about this right now.

Part of the biological argument for future discounting comes from the scarcity in which our species evolved. The more scarcity you're experiencing, the higher your discount rate will be. Because you need that $100 right fucking now. Getting $200 next month won't matter if you're already evicted ... or dead. But if you're in a place of abundance, you have the luxury of saying, sure, I'll wait a month for that extra money! Thinking about this scarcity and discounting from an evolutionary perspective, a 2013 study published in *The Royal Society*, looked at how being exposed to nature—especially abundant nature—might affect the discount rates of humans.

> Natural landscapes, especially lush ones, are intrinsically rewarding and enjoyable as they provide cues of predictability and resource abundance, at least for ancestral humans, whose psychology is likely to be still affecting modern humans. By contrast, urban landscapes—which are entirely novel on an evolutionary time scale—are inherently unstable, and convey the perception of intense social competition among humans for all kinds of resources, such as status, goods and mates. As a consequence, we hypothesize that exposure to natural scenes will make people discount the future less, whereas exposure to urban scenes will be likely to have the opposite effect.[76]

Their hypothesis was true. Even just looking at a picture of nature—never mind going on a hike or a having a picnic in the woods—was enough to alter our highly competitive discounting that makes us feel like we're constantly battling other people for resources. Of course, for most of human history, we *were* battling other people and other groups for resources.

The issue is, the most pressing problems we're facing now are all uncertain, and they're all in the future (sort of). So we just don't assign that much importance to them when we feel like we need to focus on the present. The weirdest part is we *aren't* in a time of scarcity. We're in the most abundant time in all of human history, but we still feel like we're competing for survival. Because half of us still are. And when we feel like we're in survival mode, we can't focus on anything but the most immediate threats to our existence. You can't care about the climate if you can't feed your kids.

But aren't we smart enough to do something about it now? How come we don't just all sit down at the same fucking time? Well, unfortunately, we've already created all the systems that define the way the whole thing works. We're stuck in the multi-polar trap of world domination that sees everything as a zero-sum game. If I take something, it means you have one less thing. And nobody wants less.

And because of how far down the rabbit hole of late-stage capitalism we've fallen, the people doing the plundering are the ones in charge of the only levers that could have been pulled to rein it in. Laws are now created by the businesses who stand to benefit the most from destroying the planet and using up all the "free" resources. And as they do it, we're forced to watch as the poor get poorer—as almost everyone gets poorer—and wealth is accumulated in ever more massive piles to the top 0.1%, and the wellbeing of the average American gets worse and worse despite the fact that GDP keeps on growing. All of this is just a distraction from the fact that this system hasn't worked to benefit most people in a very long time. We can't rely on actors within the system we've created to make decisions that are good for society as a whole.

And no one can pay attention to any of this because we're only programmed to react to the most immediate danger. Because the world ending in a hundred years or running out of oil or water or food in a decade isn't nearly as important as where you're getting your next meal or why minimum wage is still only $7.25 or why working a full-time job doesn't even get you above the poverty line or why Americans still don't have fucking healthcare.

Working with What We've Got

Heart, Will, and Mind

**Out of the crooked timber of humanity,
no straight thing was ever made.**

—Immanuel Kant

It may seem like everything is pretty fucking hopeless, and corporations are never going to make choices to their own detriment. And you're right. Most corporations won't ever make decisions based on anything other than profits. But people can. We have been fed this neoliberal economic narrative that everyone works solely in their own self-interest, but Smith was ultimately wrong. That's not 100 percent true. Plenty of people do altruistic things. Plenty of people don't cut down their trees to make money because they see the value in the beauty or the shade it provides. Plenty of people help other people either from the goodness of their hearts or because they feel the inextricable connection shared between all humans and all living things. Kant was wrong, too. The timber of humanity isn't crooked; the systems we've built encourage crookedness. Voltaire once declared, "We live in society; therefore nothing is truly good for us that isn't good for society."

And I think a lot of people understand that truism. Or maybe at least they're starting to. Maybe more people are seeing the effects of climate change and rampant corporate profits and ecological destruction and mass extinction and the fucked-up race to the bottom and the possibility that it's all falling apart. Maybe even more people are starting to feel those effects, and so they're even more motivated to do something—anything.

I know it feels like everyone in power is evil and people never really do good things and nothing ever changes. And you're not totally wrong; people who gravitate toward positions of power tend to be ... a little more messed up and self-interested than people who don't.

But the vicious cycle of money and power endlessly accumulating is neither inevitable nor irreversible. As former labor secretary Robert Reich explains in his salient and eye-opening book, *Saving Capitalism:*

> Equally possible is a virtuous cycle in which widely shared prosperity generates more inclusive political institutions, which in turn organize the market in ways that further broaden the gains from growth and expand opportunity. The United States and several other societies experienced something very much like this in the first three decades after World War II.[77]

We can be better. And getting better can make us *even* better. And I know another secret about human evolution. Our innate biological need for self-preservation didn't just evolve for us to kill or be killed. We also evolved to work together. Toward cooperation. People evolve individually: the most self-interested person in a group will outperform other members of a group, true. BUT, group dynamics are also a huge part of early human evolution. The group that's best at cooperation will outperform a group of self-interested people. One of the first things we started doing together as bipedal apes was stoning members of the group who weren't pulling their weight. I mean, that's a little bleak, but we've been genetically selected to work together.

There was a study done by Purdue biologist William Muir[78] who was trying to breed the most productive chickens (to make more money, of course). So he did two groups. One was a group of normal chickens, and the other was a group of "superstar" chickens who laid way more eggs than the other ones. The idea was, if I put all the superstar chickens together, they'll be an egg-producing powerhouse, and after a few generations I'll have the most productive chickens in history. Except, that's not what happened at all. The group of normal chickens—full of some chickens who were nicer, more docile, a little lazier when it came to laying eggs—did just fine. And the group of "superstar outperforming chickens" pecked each other to death.

The narrative that we as humans need to be cutthroat to survive is a false one. We need to be community-based to survive. And a society full of cut-throat individualism will only kill us all. Just like the super chickens. We need to have cooperation, and we'll find far more success. Because if we all act as though our individual needs are the only ones that matter—the way we've been going the past hundred years—we'll end up pecking each other's eyes out.

And there are plenty of humans out there who fall into the docile chicken category. There are plenty of us who want to help, who want to work together to work toward real change, but we've been fed the lie that it's impossible; that things will always be the way they are so there's no point in fighting it, and we may as well give up. But that's a bunch of bullshit. It's just that most of the people in power right now are telling us immigrants are storming the border, and woke culture wars are turning your children gay, and we have to stay busy canceling everyone who ever said anything so that we don't do anything about it. So that we don't see them lining their own pockets while they defund education to keep us stupid. So that we don't recognize the lie that the American Dream was never meant to work for most people.

Here's yet another banger from *Cultural Materialism*:

And from the general closing down of the American dream, which was founded on the rape of the previously unexploited resources of an entire continent, we get the revival of religious fundamentalism, astrology, and salvation in or from outer space. To this I would add, as a final ideological product of a decaying infrastructure, the growing commitment of the social sciences to research strategies whose function it is to mystify sociocultural phenomena by directing attention away from the etic behavioral infrastructural causes.

That is to say, people are coming up with reasons why our current broken existence isn't the fault of the system we created. A system that's now saddled with unimaginable inequality, is no longer beneficial to most of the people in it, and we're running out of resources to keep the wheels turning. Harris wrote that in 1979. And it's only gotten more true. We blame it on immigration or Trump or the decline of religion or trans people competing in college sports or whatever else, rather than the fact that the whole goddamn thing is falling apart, and most people aren't anywhere near achieving the happiness and wellbeing the system has promised since it was created.

All of those feelings I shared in the introduction—I know I'm not the only one who feels this way. There are so many people willing to work to change things. So many people who feel this frustration and want to make a difference, they've just been misdirected as to where that energy and will should go.

According to Daniel Schmachtenberger, there are three things needed to effect positive change in the world: heart, will, and mind. If you only have two of them, it doesn't work.

So if you only have heart and will, you care about the planet and you're willing to take action, but you're kind of too stupid to know what to do. So that's when you get climate activists throwing paint on priceless works of art. Hi, we see you. That just didn't help anything. Heart and will is also how we get Proud Boys and book bans and other insanely asinine responses to things that are not at all the cause of your problems. These actors truly believe they're making things better; they're just directing their energy in the wrong place.

On the other hand, if you only have heart and mind, you get a lot of intelligent yet disaffected wealthy suburban white people who probably realize shit is pretty messed up, and actually sort of care, and maybe could even make a difference if they tried (and spent some of all that money they've been accumulating), but they don't have the will to devote their time and energy and money to fixing anything. So they shrug their shoulders, say, "What difference could I make?" and go back to living their very comfortable lives.

And finally, if you have only mind and will, you absolutely can make an impact. The issue is that, if you don't have the "heart" towards doing good in the world, you're going to be motivated by selfish reasons, like greed or power. And unfortunately, the people who have will and mind outnumber the people who have all three—but only because our system was designed to maximize for money instead of for wellbeing and social good. Don't hate the player, hate the game. And that's how you get the world as it is today: the people who are smart enough to change the world, to game the system, who also have the will to do it, are in charge. According to a 2020 study by the National Institutes of Health:

> An emerging set of mainstream political attitudes—most notably radical left and alt-right, are largely being adopted by individuals high in Dark Triad traits and entitlement. Individuals high in authoritarianism—regardless of whether [they] hold politically correct or rightwing views—tend to score highly on Dark Triad and entitlement. [. . .] They're statistically more likely than average

to be higher in psychopathy, narcissism, Machiavellianism and entitlement.[79]

These are the polarizing people in our political system who are getting elected more and more. These are people with will and mind but no heart co-opting movements that started with honest people and ended with honest people getting manipulated into radical, polarizing beliefs. QAnon? Insurrection, anyone?

And they are only going to continue to make decisions that benefit themselves because that's the paradigm we were all born into. And they're going to continue to tell us it's hopeless to change it because they are well aware how powerful cooperative people can be.

So I guess the question now is, how do we find more people with heart, will, and mind to take leadership roles? How do we start to effect the changes we know we need to see from within a system built to self-destruct and built to silence anyone who threatens the existing power structure? How do we community-based folk save humanity before the self-interested chickens peck everybody's eyes out?

The Benevolent Dictator

**If we will not endure a king as a political power,
we should not endure a king over the production,
transportation, and sale of any of the necessaries of life.**

—Senator John Sherman

Obviously, "how do we save the world" is the biggest question this book (and our planet and species) is going to face. And it's not easy, and it's certainly not simple, and I certainly don't have an answer. But the ideas I do have are both top-down and bottom-up. We look for small, individual changes and massive political and societal changes. We look for any levers we can pull big or small to get more people working toward a society based on human wellbeing that is capable of living in harmony with the only planet we've got. We look for people with heart and will and help them figure out the 'mind' part. We look for people with heart and mind and help them figure out the 'will' part.

We need to start thinking about every piece of our society that is benefiting corporations at the expense of the planet or at the expense of human wellbeing. Because those are the only two things we should be optimizing for. Humans will always make markets—it's been that way since we first encountered other humans and started trading with them. But since we know that late-stage capitalism in a largely unregulated global marketplace searching for infinite growth is bad for humans, how can we shift it to be better for more humans? How do we start to pull some levers to create a system that recognizes the limits of the ecosystem we are part of while still maximizing wellbeing for the most people?

Since we don't naturally have these checks and balances of insular social communities anymore, we need to create them. We need to make rules that force society to start making decisions for the good of society. We need a benevolent dictator. So here are all the laws I'd immediately sign within the confines of the system we've got. It's my benevolent dictatorship for a day.

Ranked-choice voting.

I want to start with a couple of easy ones that already have relatively broad support and already exist in a few states. Ranked-choice voting means when you go to vote, you pick both your #1 choice, your #2 choice, and so on. If your #1 guy doesn't get it, then your vote counts toward #2 (instead of not counting at all). It's sort of like an instant run-off election, where the guy with the least votes is disqualified, and the votes are re-tallied allocating the votes for that guy (or gal or non-binary human) to their second choice. The reason this is a good thing is because it decreases polarization. Candidates are forced to pander not just to their (increasingly) extremist base, but also to people who might consider putting them as their number two. Considering political polarization is one of the biggest issues stopping us from getting anything done when it comes to corporate greed, late-stage capitalism, the climate—everything—this is an easy way to force candidates to walk back the extremism.

Now, the argument can be made that ranked-choice voting has the ability to increase the electability of extremists, as extremist candidates will more often be the first choice, and the moderate third-party candidate will get dropped, leaving only two, equally extreme options to go head-to-head. And the one who had more of the moderate vote as their second choice would become the winner. But there are plenty of different "rules" for tallying votes within a ranked-choice system that can be used to avoid this outcome in extremely polarized electorates.

Using the Borda rule, for example, each of three candidates would be assigned points: 2 points to a voter's top-rated candidate, 1 point for the second choice, and 0 points for the third person. In this system, if both extreme left and extreme right chose the moderate person as their second choice, we'd get a moderate candidate elected. I'm not assigning any specific value to a moderate candidate over an extremist one, but looking instead to incentivize candidates to appeal to a broader base of what people actually want in the first place. If anything, I lean pretty far left, but we need to reduce overall polarization if we have any hope of getting anything done.

Get rid of gerrymandering, obvi.

I mean, come aaahhhhhhnnnnn. How is this still a thing? In case you don't know, every ten years when the US Census Bureau releases their new census data, political districts get redrawn due to changes in population. Makes sense. Except, the thing is, the party in power is the one who gets to do the redrawing. So, what do you think they do? Obviously, they draw the new districts in a way that benefits their own party the most, at the expense of what the population—the actual voters—want.

There are two types of gerrymandering: "packing" and "cracking." Packing is when they draw super-weird lines around counties to "contain" the portions of the population belonging to the other party, giving them fewer districts overall where they hold a majority. Their candidate will definitely win in that district (since it's now like 90 percent Democrat), but there won't be enough of them in any other districts.

"Cracking" is basically the opposite. It's when they split up members of a party (or race or socio-economic status) so they can't make up a big enough piece of the pie to get their candidates elected in *any* district. Since large cities tend to be more Democratic, you can pack those districts as tightly as possible, and then "crack" the remaining Democratic pockets across the rest of a state.

You can look at this lovely map of North Carolina's 12th district to see how they've stretched that Democratic pocket from Charlotte all the way up to Winston-Salem, ensuring that two Democratic strongholds walk away with only a single congressional seat.

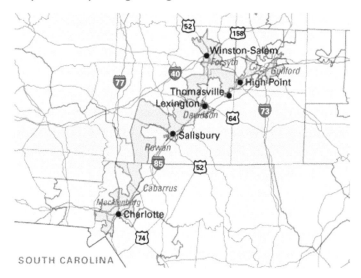

Nationally, gerrymandered congressional maps gave Republicans a sixteen- to seventeen-seat advantage for most of the last decade since the 2010 census. And on a state level, it's no better. In 2018, Democrats in Wisconsin won every statewide office and a majority of the statewide vote, but somehow only ended up with thirty-six of the ninety-nine seats in the state assembly.[80]

I can't believe I have to say this, but: political parties should not be in charge of drawing the maps that decide which political party will win. And while the Republican party perhaps has some more egregious examples of gerrymandered districts in the past decades, Democrats are guilty too. No matter which party is doing it, it's the will of the people that loses out. Wild idea: how about non-partisan committees redraw the district lines every decade? Fuck it, how about you just have the Census Bureau do it? What if politicians actually cared about what people wanted instead of just about winning? Lolz.

Overturn Citizens United.

If you already knew about gerrymandering, then you've probably heard of Citizens United. Citizens United was a landmark 2010 Supreme Court decision in the case of *Citizens United v. the Federal Election Commission*. Essentially, a conservative non-profit organization wanted to air a hit piece about Hillary Clinton just before the 2008 primary elections. But we had a law against corporations, non-profits, or unions making any "electioneering communication" within thirty days of a primary or sixty days of an election. The case made its way to the Supreme Court, and they decided this was a violation of free speech. They basically said, "Businesses are just people, and we can't tell people what to say and when because of the First Amendment." In the same decision, they also overruled *Austin v. Michigan Chamber of Commerce*, which both prohibited election spending by incorporated entities (read: businesses) and limited the amount of money businesses could spend on elections.

In case you're not following, this decision made it possible for corporations of any size to spend as much money as they want towards electing whomever they see fit. It led to the creation of Super PACs, which are multi-million-dollar slush funds for politicians where corporations can put in as much money as they feel like. They're not donating directly to the politicians (which is still highly regulated), but those super PACs can spend as much money as they want running as many ads as they want, whenever they want. It's no surprise that the candidates they are supporting are both aware of and in favor of these super PACs. Considering political campaigns spend basically all of their money on media, this decision created bottomless corporate donations to the candidates of their choosing.

The end result? More power to large corporations and lobbyists. More corruption. More dark money. Think about the oil and gas companies "donating" tens of millions of dollars to politicians they know will vote against any climate-related bills. Think about big banks spending tens of millions of dollars to help elect politicians who will keep financial regulations loose so they can keep up their patterns of reckless lending and negative pig creation. The implications are dire and far reaching. Massive corporations aren't people. And they shouldn't get a vote in the future of our country because they don't give a fuck about the future of our country.

In case you can't fully grasp just how much control corporate interests have in our political system (and therefore our lives and financial and emotional wellbeing), two Princeton professors did an analysis back in 2014 of the relative influence of economic elites, business groups, mass-based interest groups like unions, and average citizens across 1,799 pieces of legislation.[81] You know what they found? "The preferences of the average American appear to have only a minuscule, near-zero, statistically non-significant impact upon public policy."

Woof. Even I didn't think it was that bad.

No private funding for elections.

Considering how bad it's gotten, how about we just take the money out of politics altogether? We could have publicly funded elections, and each person gets the same amount of money once they receive enough signatures to join the ballot? They can use that money to throw fundraisers or print yard signs, whatever. But there would be no political advertising permitted on television or social media, and the winning candidate wouldn't be the guy who has the most money or who tells the best lies. People would maybe have to—get this—go learn about candidates on their own??? Okay, that one's not gonna happen. Let's redirect our attention a bit from politics to big business.

Ban advertising.

If we're banning political advertising, fuck it, let's just ban the whole damn industry! We may all think that humans just really love buying stuff. But if we loved buying stuff so much, then why do corporations have to spend billions and billions of dollars convincing us to buy stuff? Back in the 1970s, it was estimated the average American saw 500 to 1,600 ads per day. Marketing firms were still figuring out how to push our buttons, but the ads were mostly confined to magazines and billboards, and of course, television.

If you grew up in the '70s or '80s or even '90s, you probably started to get that feeling that your shows were getting shorter and the ads were getting longer. They were. A 30-minute episode of *The Simpsons* in 1989 was 23 minutes long. By 2017, the run-time had dropped to 21.3 minutes.[82] All in the name of ad revenue. And when they couldn't squeeze any more ads into the commercial breaks (because there is a human limit to what we will tolerate), they started to put those little pop-ups on the bottom of the screen *during* the show you were watching.

And then, we got the internet. In 2007, a market research firm estimated we were each seeing up to 5,000 ads per day. Fast forward another fourteen years to 2021 and that number has grown to 6,000 to 10,000 ads . . . every. single. day. Google alone made $134 billion in advertising revenue, just in 2019. There is now a Russian startup that wants to project advertisements INTO THE FUCKING SKY AT NIGHT. Is there no limit?*

Sure, you probably agree with banning advertisements in the night sky, or on the Grand Tetons or whatever, but otherwise it probably sounds pretty radical . . . except . . . it's not. If we banned advertising for cigarettes and alcohol and most other countries have banned advertising for prescription drugs (seriously, why do we have ads for Prozac in America?) then why not just ban it all? Sweden banned all advertising to children under twelve. Prescription drug ads were already banned in the US before Big Pharma insisted otherwise. How about only print advertising is allowed, like for newspapers and magazines? What's that you say? This will collapse a massive chunk of the economy? Millions of websites won't be able to operate without all the money they're raking in from Google ads and affiliate marketing? You mean I won't be inundated with shitty clickbait articles written by ChatGPT? OKAY. That's kind of what we're going for here.

Back in the '70s, they may not have been great at understanding how to use advertising to manipulate our tiny caveman brains, but now they're really, really good at it. They know exactly how to make you feel.

---

* If this ever happens, just kill me.

They know how to make you feel inferior, insecure, and how to make you feel like their product will solve all of those things. Ad companies literally hire scientists to do psychometric research into how to better manipulate you.

All advertising does is convince you to buy things you don't need or want. It's driving that wonderful growth every business is looking for at the expense of our planet, the expense of our personal financial situations, all mostly to the benefit of massive corporations. The more you start to think about the parts of our daily lives that don't benefit the majority of people, the easier it is to start thinking about a world without them.

If you say, "Well, my small business couldn't survive without advertising!" Then okay, let's make a rule for companies over a certain top line? Companies with over $40 million in sales can't advertise? That should get rid of a pretty big chunk. I don't think anyone ever really needs to see an ad for Coca-Cola or fucking Exxon.

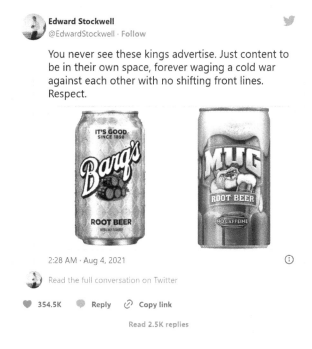

What if we just bought the things that popped into our heads when we needed them? You could still browse your favorite online stores or walk down a shopping strip lined with overpriced boutiques. There just wouldn't be ten thousand Instagram posts making you feel like shit for *not* having those things. Would that be so bad?

Actually, let's just ban a whole bunch of things.

Advertising as an industry is bad for society. So, we ban it. What else can we ban on this premise? How about defense contractors? Private prisons? Lobbyists? Planned obsolescence? Fast fashion? This is getting fun!

For everything we take as a given part of our daily lives, we need to ask ourselves if it's advancing our wellbeing as a society. Now, we're obviously not going to get rid of the military (even though we should definitely unwind it by at least half), but there are a lot more things we can easily prohibit within the confines of our current system. Let's start looking at things through the eyes of the benefit of society and not through the "rights" of corporations to harm us. Any industry that at its core is in the business of human health or wellbeing should not be able to operate privately for profit. This includes healthcare, prisons, and weapons manufacturing for sure. Why can't they be private? Because they are perversely incentivized to maximize profits to the detriment of human lives.

Private prisons shove more and more human beings into smaller spaces, feeding them worse and worse food because they have a responsibility to their shareholders to maximize profit. They pay off judges to give stricter sentences to make sure their prisons stay full. They lobby for stricter minimum sentencing requirements to make sure as many people in this country go to jail as possible and stay there for as long as possible. They legally HAVE to do anything they can to be as profitable as possible. If they make choices that are, say, better for human lives but worse for business, they could be sued by their shareholders. Okay, I've got another idea.

Update fiduciary duty for corporations, aka Stakeholder Capitalism.

As it stands, any publicly traded company in the United States has a fiduciary duty to its shareholders. The way these things are worded are the "duty of care" and the "duty of loyalty." Well, that sounds all warm and fuzzy! This means the company must act in the best interests of the shareholders when making decisions. The problem here is that operating in the "best interests" of shareholders is, 100 percent of the time, making more money. Maximizing profits. This is known as shareholder primacy. They come before everything and everyone.

But what the hell do shareholders have to do with anything? Let's think about a car manufacturing plant in Detroit. The plant is run by say, a thousand workers who come in every day. The plant provides a lot of jobs for the town, but it also spews out a bunch of carbon dioxide into the town's air and dumps its toxic byproducts into the town's water supply. The people who are affected by that company's behavior aren't the shareholders—it's the stakeholders. When they cut wages or make massive layoffs, it's the stakeholders who get hurt. When they dump their toxic waste in the river, everyone in the town suffers, but the shareholders are happy because the CEO increased quarterly profits by not paying to properly dispose of their waste.

There is a movement known as the B Corporation[83] (or Benefit Corporation), where businesses voluntarily elect to operate on stakeholder primacy. Stakeholder primacy is where a business has to consider the interests of not just shareholders, but *stake*holders: customers, workers, suppliers, communities, investors, and the environment. What decisions would be best for all of those things?

Fifty-one jurisdictions around the world (forty-four of which are US states) currently have stakeholder governance as an option for companies, but it's still not a requirement.

So let's make it one. No company should be able to operate without considering the impact they're going to have on the people who will actually be impacted.

I recently learned that CEOs used to do this voluntarily, seeing themselves as "corporate statesman" responsible for the welfare of the communities they serve. In another eye-opening passage from *Saving Capitalism*:

> In the first three decades following World War II, corporate managers saw their job as balancing the claims of investors, employees, consumers, and the public at large. The large corporation was in effect "owned" by everyone with a stake in how it performed. The notion that only shareholders count emerged from a period in the 1980s when corporate raiders demanded managers sell off "underperforming" assets, close factories, take on more debt, and fire employees in order to maximize shareholder returns.
>
> In November 1956, *Time* magazine noted that business leaders were willing to "judge their actions, not only from the standpoint of profit and loss" in their financial results "but of profit and loss to the community." General Electric famously sought to serve the "balanced best interests" of all its stakeholders. Pulp and paper executive J.D. Zellerbach told *Time* that "the majority of Americans support private enterprise, not as a God-given right but as the best practical means of conducting business in a free society. [. . .] They regard business management as a stewardship, and they expect it to operate the economy as a public trust for the benefit of all the people.

Fast forward a decade, any CEO who cared more about people than profits was ousted, Milton Friedman told stakeholders to go fuck themselves, and here we are.

Employee ownership capitalism.

And that ties in just perfectly with my next idea, one that has been gaining traction in recent years: employee ownership. Right now, we have shareholder capitalism. A company is owned by the people who started it (and probably a bunch of venture capitalists) until they go public, at which point they become owned by their shareholders. Of course, the owners usually hold on to a healthy portion of their own stock. All the profit that public companies make, all the value they create, goes to shareholders (including themselves). As we just discussed, public companies are legally required to make business decisions with shareholders' best interests in mind. And when the company makes a billion-dollar profit, the employees who made that profit possible through their own skills and value and labor see none of the upside. And when that upside was created by downsizing the staff or cutting their pay, the employees keeping the whole thing running probably aren't too thrilled about seeing the stock go up by three points. But when employees have joint ownership in a company, they are strongly incentivized to make decisions that don't negatively affect . . . themselves.

So let's make sure that any company over a certain size is required to have 49 percent employee ownership. You give up your equity and transfer ownership to your employees. If you're publicly traded, half of your shares never get put on the market. Plenty of businesses offer some kind of profit-sharing for employees. We're just going to make it required once you get too big. This one actually seems pretty simple to me, other than forcing publicly traded companies to buy back 49 percent of their stock. But we'll figure it out. They LOVE doing stock buybacks to drive up their share prices. In this world, we just force them to transfer those shares to employees. Ooh! And in addition to employee ownership, you could take it one step further: Union leaders could be required to have a seat on the board; mayors of cities where the business operates could get a vote as well. There are lots of options here! Stakeholder decision-making for the win!

Okay, back to banning shit.

As you start reading some more of these ideas, you'll start to see more clearly why Citizens United needs to be overturned. Because no politician is going to vote for a bill that would hurt the corporations who are funneling tens of millions of dollars into their campaigns.

And if any bill gets introduced that would hurt any specific corporation, you can bet they're gonna spend as much money as it takes to make absolutely sure that shit won't pass.

Which is why I would like to ban: **lobbying!** Sure, lobbying can be a good thing when people are lobbying for causes that will benefit more people. Or save our planet. Or stop an oil pipeline from destroying sacred indigenous lands. Or lobbying for any of the things I'm adding to this list. But those noble causes will never have as much money as the corporate overlords. So they will *always* get out-lobbied. Lobbying has grown from a $1.45-billion-dollar industry in 1998 to a $3.73-billion-dollar industry in 2021.[84] Where is all that money going? For one, it's going from Big Pharma into the pockets of politicians to make sure we don't cap prescription drug prices or to block universal health care.[†] It's going from private prisons to politicians to make sure we don't reduce minimum sentencing times for criminals. And it's a lot of smaller things in your life you may never think about. Cable companies in America have successfully lobbied twenty states to enact laws prohibiting their own cities from laying fiber cables. Why on earth would a state ban its own cities from improving their own infrastructure? Infrastructure that arguably could increase your state's GDP and job growth and other amazing things? That's the power of deep pockets, guys. And they keep paying those politicians to make sure they continue to toe the line.

Feel like the cost of rent has gone absolutely bonkers in your city or town? Well, private equity firm Blackstone may have had something to do with it. Their current business model involves buying as much residential property as possible, jacking up the rents as high as possible, evicting existing tenants who can't pay, and basically just rinse and repeat. This single firm owns more than $326 million worth of residential real estate in the country. But since some cities had pesky rent control laws that were stopping them from jacking the rent up 20 percent, it was in their best interest to fight them. So that's what they did and do. In California alone, "Big Real Estate" (yes, apparently every industry now has a "Big"

---

† Big Pharma by far outspends any other industry with $373.7 million spent on lobbying in 2022 alone.

in front of it) spent $77.3 million to stop a single rent control bill from passing. And it's not just at the state level. They lobby towns and cities to keep regulations loose and hold onto their ability to arbitrarily raise your rent while bragging about "healthy cash flow increases" on investor calls.

So, what's the solution? Well, just ban lobbying. Ban lobbyists. Ban being able to be paid to go bug politicians.‡ Because we all know there's a whole bunch of quid pro quos, pocket lining, you-scratch-my-back-I'll-scratch-yours going on over there. And we also have to ban the practice of politicians moving right from their cushy jobs in Congress to cushy jobs lobbying for those same companies as soon as they're done. This practice is so common it has a moniker: the revolving door, where politicians go to work for high-level positions in the private sector and vice versa.

In the 1970s, only about 3 percent of retiring members of Congress went on to become Washington lobbyists. But now? Fully half of all the retired members of the 115th Congress (2017–2018) became lobbyists, regardless of party affiliation.[85]

The Close the Revolving Door Act of 2019 aimed to ban this, but guess what? It never even made it to the Senate floor. Shocker.

Moving on: **fast fashion.** This is absolutely terrible for the environment. But how do you ban it, exactly? It's just cheap fabrics sewn cheaply into cheap clothes using cheap labor (uh-hem, slave labor) in countries that are cheaper than ours. Then those cheap clothes get shipped across the ocean using super cheap shipping and BAM you've got a shirt for $2.99 that you'll wear twice before it falls apart, or you just get tired of it. Well . . . I guess we could start by banning the import of anything made in hazardous or unsafe working conditions? This is tough. Maybe tax the shit out of it so it's not so cheap? Or maybe everyone just agrees to

---

‡   And maybe also just ban private equity firms from owning residential real estate in the first place?

stop fucking buying it? Oh! I know! We create minimum wage requirements for any products being imported. You wanna get on the list of
approved importers? You've gotta pay a minimum wage at least half of
the median income in the country where it's made. Is that even possible?
I don't know. I'm open to ideas here. The EU is working on a law[86]
setting a mandatory minimum percentage for recycled fiber in clothing
and would ban the destruction of unsold products (which companies do
because it's cheaper to just throw them out than do literally anything
else with them). So, you know, some things are happening!

Speaking of designing clothes knowing that they're going to fall
apart after a few wears, let's talk about **planned obsolescence**. If you're
not aware, this is the practice where companies make things they know
will break so that you have to buy a new one sooner rather than never.
It's like how for some reason your phone always stops working after
three years, but that fridge your parents bought in the '70s will outlive
your children and your children's children.

Back in the 1950s,
America had sort of al-
ready reached that sec-
ular stagnation that
Smith and Mill were
so pumped about.
Consumption had pla-
teaued. People had all
the things they wanted.
And businesses start-
ed freaking out. If peo-
ple weren't going to
keep buying things, the
whole model falls apart!
*How can we get them to
keep buying the things???*
So one day—because of
this necessity for endless
growth—companies

2022 appliances: *break within 2
years*

1970s refrigerator: I will outlive you
and everyone you love. I am eternal.
I am time itself

decided that if everybody already owned a fridge, and fridges lasted for fifty years, it was probably better to start making fridges that broke instead of selling fewer fridges. And thus, we have planned obsolescence. How do we ban it? It's not super straightforward. There are right-to-repair laws popping up, meaning that Apple or John Deere can't void your warranty if you try to fix a product you own. There could be massive—and I mean MASSIVE—fines if a company is discovered to be engaging in planned obsolescence practices. There could be required guarantees on all products based on expected lifespans set by an industry watchdog group and approved by Congress. We all believe a fridge should last twenty years. So if your fridge doesn't, the company is forced to replace it. This would weed out not only planned obsolescence, but businesses that just make shitty products and don't deserve to be in business.

Anyone remember how L.L. Bean used to have a lifetime guarantee on EVERYTHING? And REI as well on all of their REI-branded products? That shit was amazing. And people loved and trusted those brands and those companies didn't go out of business. They got loyal followers who would buy everything they needed from those stores. Until some new CEO came in and was like, "Hey, this is costing us money, have we thought about just *not* doing that? And now your $200 boots have a one-year warranty instead of a lifetime warranty, and they're probably made with worse materials in a factory in China staffed with nine-year-olds. Le sigh.

In case banning planned obsolescence sounds impossible, France already did it in 2015, and they're just now starting to see lawsuits against companies like Apple for building their products with intentionally shorter lifespans. In theory, if these lawsuits are successful, it will make other companies think twice before they engage in the same practices. Will it work to stop the practice altogether? I don't know, but I love seeing countries test out these ideas! And once again, the EU is already working on a broader plan as well, forcing companies to provide information on a product's expected lifespan, along with banning any "eco-friendly" claims that are just totally made up. And the penalties? Up to 4 percent of a business's annual profit and even confiscation of ANY profits related to that product. Dang EU, good for you.

These are things that people want. Consumers want to know how long their fridge is expected to last. Consumers are willing to pay more for products they know won't break. And businesses that make great products get to benefit as well. There is literally no reason not to be doing this.

Overhaul patent law.

This may sound random to you, but patents are one of the basic rules of a market that must be established: who owns what for how long? If I invent something, do I own the idea forever? Or does it eventually become part of the public domain? When the Founding Fathers wrote the Constitution, they realized this was an important thing to address and even included language authorizing Congress to "grant patents and copyrights to promote the progress of science and useful arts by securing for limited times to authors and inventors the exclusive right to their respective writings and discoveries."

But did you notice that wording? "Securing for limited times." Because owning a patent forever is bad for the common good. Science and art and society are better off when inventions and ideas can be shared and improved upon. When one company owns a patent for a technology that no one else can use or compete with, they have no incentive to ever improve that technology!

Right now, patents in the US last for twenty years. That sounds reasonable, I guess. But companies can change one tiiiiiny detail in their design in order to get the patent renewed for an extra twenty years. And they can keep doing that forever.

Also, a lot of companies will buy patents they see as a threat to their business model . . . and just bury them. They don't use them; they just don't want the competition out there. Obviously, this is very un-capitalistic of them. So let's make a law that if you file a patent, you have to do something with it in like . . . five years. Or at least show you've been working towards something. And if you haven't? The patent becomes public domain, and anyone can use it to develop anything they want. There could be some amazing advancements in say, clean energy, that the fossil fuel industry might buy just so they can be sure they never see the light of day. Because they don't want anyone hurting their very stupid and short-sighted business model. There are some fun conspiracy theories out there that we already had a car that could run on just water, but the oil companies bought the patent, killed the guy, and here we are. It's 100 percent not true, by the way. You can read about Stanley Meyer's fake water fuel cell car on Wikipedia. Or maybe that's what the oil tycoons want you to believe . . . JK.

And patent reform needs to go much further. Right now, plenty of Big Pharma companies operate on a pay-to-delay system. Once their patents expire, they simply pay the makers of generic drugs to delay release of the cheaper versions. The generic drug company makes money, the Big Pharma company can still charge whatever they hell they want since they have the only drug in the market, and it's all completely legal. And who gets fleeced? You do!

If you think these pay-for-delay agreements are normal, they're not. They're banned basically everywhere except the United States. And plenty of countries also have laws banning patent extension by changing insignificant details. It's not that radical, guys.

Ban billionaires!

Speaking of rampant corporate profits at the expense of everyone and everything, I've got another good thing to ban . . . BILLIONAIRES! Any money any individual makes over $1,000,000,000 can either be taxed at 100 percent, transferred to employees as profit-sharing, or treated as required philanthropy. Sounds complex? Yeah, I know it is. But arguments saying "this would never work" or "billionaires would just move their money overseas" are lacking in imagination. First of all, we know who these motherfuckers are. Ain't no billionaires hiding in the world. They use loopholes to take out loans against the value of the shares of their stock that they basically never have to pay back until they die. First, ban that practice.§ Force them to sell stock in order to

---

§   Man, banning things sure is fun!

buy things. Or force them to pay themselves a salary that represents their actual "incomes," which they would then have to pay income taxes on like everybody else. The IRS already has laws requiring "reasonable" salaries for normal people who pay themselves out of their business profits. So devote a small team at the IRS to individually deciding the "reasonable" salary of each billionaire (it's not like there's that many) . . . and then tax the ever-loving shit out of them.

There are SO many ways to avoid paying income tax as a wealthy person. I don't know them all because I'm not one, but trust me—we could probably write a whole book just on that.

Second, literally just base the threshold on the net worth of the company. If a single shareholder's ownership in a company becomes worth more than $1,000,000,000, then they have to sell or transfer shares in order to get their net worth back below a billion. If they sell it, they have to give the proceeds from those sales to the government or to employees or to a charity. Can it be their own charity? I guess so, but we'll need some serious oversight there.

I know there are a lot more complexities here about how the stock market works and equity dilution and some other shit. I'm open to whatever ideas you've got to stop billionaires from existing. I don't give a shit if the markets won't like it or billionaires won't like it. This is my benevolent dictatorship, so deal with it.

Cap CEO income.

A handy way to maybe just stop billionaires before they start is to . . . stop paying CEOs hundreds of millions of dollars? A CEO of a company should not make 300× what their average employee makes. This may come as absolutely zero of a shock, but back in the '50s, the CEO-to-worker pay ratio was just 20-to-1. That seems pretty reasonable, I guess. You make $50,000 and the CEO makes $1,000,000. Okay. But then . . . shit got a little out of control. In 2019, that number was 264-to-1.[87] In 2020? 299-to-1. And in 2021? That's right: 324-to-1. That means you make $50,000, and your boss makes *checks notes* $16.2 million a year. That's $1.35 million per month. Or $67,500 PER

DAY (assuming twenty working days per month). And that's just *an average*. Wanna guess Google's CEO-to-worker ratio? Over 800-to-1. And Apple? Almost 1,200-to-1.

You wanna know the only reason we haven't made this a law already? Well duh, because the people running corporations are the ones who decide which laws get passed. And I don't think they'd be very big fans of this one. And something about free-markets, yada yada yada.

"But Taylor, if you cap CEO pay at only a million dollars a year, no one will be incentivized to invent new things! Everyone will just sit around on their asses being uninventive because who would ever work for only a million dollars a year?!"

Well, I won't bother responding to how ridiculous that argument is. But I don't have to . . . because it's a RATIO. If we capped it at even just 50×, you could make $20 million a year . . . as long as your employees average out to $100,000. Let's take a look at a non-famous CEO (as in not Elon Musk or Tim Cook or someone you've heard of). Let's look at David Baszucki, the CEO of Roblox. I had never heard of him until five minutes ago, but according to an annual study conducted by Equilar, he is the seventh-highest paid CEO in the United States. In 2022, he made a paltry $232,786,391. That's $232.8 million dollars. In a single year. Any guess what percent his pay raise was over 2021? More than three thousand percent! Man, he must be really good at CEO-ing.

Now let's take a look at the company. Even though Roblox is a "gaming company," they don't actually make any video games. They are just a platform where you can play games *other* people made. So they literally **don't make anything**. Roblox's fiscal year ends in September, and their revenue for the twelve months ending in September 2022 was $2.251 billion, a 33.3 percent increase over 2021. Well done, David Baszucki. You brought in over two billion dollars without making a single fucking video game.

According to the website comparably.com, the average compensation at Roblox is $134,365 annually. That's pretty respectable, but I don't see any data in here for janitors. The lowest paid job listed here is administrative assistant at $56,853. Presumably, as a tech company, most of their employees are developers and engineers who get paid pretty well already. Roblox employs about 1,600 people, so if we multiply $134,365 × 1,600 employees, we get $215 million. So Baszucki already makes more than ALL OF HIS EMPLOYEES COMBINED. If we add their salaries to his salary, we get $447.8 million.

Now, if the CEO salary were capped at say, 50× the average, Baszucki could still be making $6.7 million a year without giving any of his employees a raise. Doesn't sound like enough? Well, you can keep giving yourself a raise, so long as everyone else gets one too. Isn't this fun! Baszucki could make $12 million a year while more than doubling the average employee salary to $272,000 . . . and still have a lil' money left over!

But if we want to get real crazy, we can cap CEO pay at 20×—or even 5× or 10×—the average worker. In fact, there's already a nonprofit group called WageMark based out of Canada that's fighting for an 8× cap on CEO salaries. This is NOT a radical idea.

In Australia, shareholders can force an entire corporate board to stand for re-election if 25 percent of shareholders vote against a CEO pay plan two years in a row. And it works! In 2013, CEO pay in Australia averaged just 70× the average worker, compared to our absolutely insane 324× multiple.

To be fair to America, some people are already working on this here. The California state legislature introduced a bill in 2014 that would have introduced a sliding tax scale based on the CEO-to-median-worker ratio. If a CEO earns 100× as much, their tax rate drops from the 8.8 percent corporate tax rate to just 8 percent. If they make just 25× as much as their typical worker, it drops to seven. And if the CEO earns 200×? They pay 9.5 percent; 400×, 13 percent. And then of course, right-leaning, business-owned media outlets start screaming about how this is gonna hurt the economy. No, I assure you, it will not.

Trickle-down economics is the biggest joke in history. It never worked (except that one time) and it never will again. Businesses don't reinvest their profits; they use them for stock buybacks to increase share-holder value and pay their CEOs even more money that is conveniently not taxed as income. The bottom line is, nobody needs to or should make more than $100 million dollars a year.

Tie minimum wage to inflation.

While we're talking about billionaires who definitely became billionaires on the backs of thousands of minimum-wage workers (because there is literally no other way to do it), here's another layup. Tie minimum wage to inflation instead of having it voted on by a partisan government body beholden to corporate interests. This one also is decidedly *un*radical. It's just obvious. There are dozens of countries who already do this. In some cases, minimum wage isn't decided by any humans at all. Within the EU, there is some form of automatic price indexation of private sector wages in Belgium, Spain, France, Cyprus, Luxembourg, Malta, and Slovenia.

Arguments against this say with rising inflation, rising wages could cause shock to businesses—forcing them to close when they can't afford these newer, higher wages. Oh no! I say stop protecting any businesses over a certain size. Sink or swim, motherfucker. You can't afford to pay your workers a living wage? Then congratulations, you have an unviable business model! Wanna protect small businesses? Have a different minimum wage for companies that employ under a hundred people and over a hundred people. That California CEO pay bill only applied to companies with over $10 million in revenue. Have an entirely different set of rules for companies that employ over a hundred people. Incentivize businesses to stay small because it's literally better for everybody. Have a non-partisan committee of economists vote what the minimum wage increase should be every year because let's not split hairs here—if the economy grows every year, and inflation grows every year, and the cost of groceries and rent and clothes grows every year—so should the minimum wage. If you don't get a raise every single year, you're getting a pay cut. And I'm pretty sure we all got some pretty massive pay cuts in 2021.

Tie Congressional salaries to median household income.

Ooh, this one is super fun! And I'm killing these segues right now. Think it's easy to live in America on the policies you've created? Okay, then give it a whirl! Can someone explain to me why senators make

$172,000 a year when the median salary for an individual is less than $40,000?

It's very important here that we do median and not average. Because those CEOs making $300 million a year are really pulling that average up. Median just means right in the center of all the numbers. To put it into context, the *average* household income in 2021 was $97,962. But the median household income was just $69,717. Keep in mind—that's household—not single earner. These congresspeople are probably in dual-income households.

This is all a game about incentivization. It turns out, while some people are good, lots of people suck. Because our society values money above all else, most people make most of their decisions in line with what makes them the most money. Since shifting the entire value system of our culture is difficult (though I'm still gonna talk about doing that later), for the time being, we have to put policies in place that use the system we have to incentivize the behavior we want. A policy like this would force Congress to make decisions that would increase the median income in the country. Like increasing minimum wage or operating on stakeholder capitalism. What's good for the goose is good for the gander, that sort of thing. Right now, our policy-making situation is all goose and no gander.

Sliding-scale socialism.

While we're busy requiring businesses over a certain size to be 49 percent owned by actual stakeholders, let's just put a cap on how big privately-run companies can be. I mean, think about it. When a company gets so big that it affects the lives of everyone in a country—think Coca-Cola, Verizon, Walmart, fucking Nestle—they shouldn't be able to make decisions that will harm that many people. And the profits they make are coming at the expense of hundreds of millions, even billions of people for companies like Coca-Cola. So the larger a company gets, the less ownership a handful of white dudes get to control. Over a certain size, the company—and its profits—would become public goods (or maybe some mix of public and employee ownership).

Now, I know socialism is a big, scary word. I also know most people don't even know what it means. It means that the ownership of the means of production (all those businesses and their billionaire CEOs) are owned by the community as a whole. The problem is, that's never actually happened in history. Venezuela is not socialism (their oil production is state-owned, but they also have lots of capitalism and dictatorship and corruption and failed economic policy mixed in there). Nazi Germany was not socialism (despite being called the 'National Socialist' party in an attempt to appeal to the suffering lower middle class). And China most certainly is not. The closest thing I think we have is Scandinavia (Sweden, Norway, and Denmark), which most people refer to as "Democratic Socialism." You still have private enterprise, but with broad social safety nets and massive government-run programs protecting the welfare of the people and public ownership of things like oil. In Norway, pretty much everything that's in the land is owned collectively by the people. There is a state-controlled "oil pension fund" and every Norwegian owns an equal share. Right now, that fund is worth $1.4 trillion dollars. That nearly $260,000 for each and every Norwegian. Including babies. That's money that would have gone straight to an oil and gas company in the States. And oh my God! Their whole society hasn't collapsed like Venezuela! How did they do it?

Democratic socialism means you can have both. Lots of big industries like energy and healthcare are public, but you can still own a restaurant or make money inventing a cool new technology. If you want to call taxpayer-funded public healthcare and public school and public road building socialist, then okay, I guess I'm a socialist.

This one is probably going to upset the free-market capitalists the most. But I'd also challenge anyone to think of a country with NO public goods (I'm looking at you libertarians). Who builds the roads? Who takes out the trash? What about socialist libraries? And socialist fire engines? The common argument from most free-market capitalists is that private enterprise does it better and cheaper. You're not wrong. Government-run programs are notorious for being slow, bloated, and wasteful. But when the choice is between "slow and bloated" or "this corporation is destroying our planet and our community on the backs of hundreds of thousands of humans whom they systemically exploit and underpay," I'll take slow and bloated any day of the week.

And of course, it's unlikely this would ever happen. Because as soon as you set the cap over which a business can't grow, business owners will break apart their own businesses to stay under that cap. This would, once again, incentivize companies to stay smaller so that the owners of those companies could pocket more of the profits. And since smaller companies are better for everyone, it's a win-win!

Make companies pay for all waste produced and incentivize closed-loop systems. Make companies pay.

On the topic of corporations being responsible for most of the destruction of the planet, here is another idea to help them stop doing that. Once again, we're back to incentives. Companies don't have to pay for the waste they produce. And thanks to some clever ad campaigns paid for by these companies in the 1980s, consumers have been led to believe it's all our fault, and if we just recycle, everything will be fine.

"Keep America Beautiful, Inc," the organization that paid for the famous "Crying Indian" PSA, is actually an organization founded by leaders of major packaging and beverage corporations like Coca-Cola and the Dixie Cup company meant to shift the public conversation away from how much waste these companies were producing once the industry shifted to single-use plastics and away from refillable bottles. Also, the dude in the ad is an Italian American, because of course he is.

So let's force companies to pay for the waste they produce. We could tax new, single-use plastics to oblivion, forcing recycled plastic to become cheaper than new. Why would anyone pay extra money for recycled plastic when we have unlimited petroleum to just keep churning out 1.3 billion plastic water bottles each year? Amirite?

Again, this isn't a totally new idea. Some countries are starting to tax corporations for the amount of waste they produce. The EU recently introduced a "plastics tax" as part of their EU Green Deal, which is up to each country to structure. In the UK, they're charging companies £200 per ton of plastic packaging components (that is very little money, btw). Spain has introduced a tax of €0.45 per kilogram only on non-reusable plastics (also very little money).

This is nowhere near enough. Fining multi-billion-dollar companies doesn't do shit when the fine is less than the cost of making the change. They'll just keep on breaking the rules and paying the fines because it's cheaper than changing their behavior.

Most of the "fines" corporations get slapped with for breaking laws equate to a fraction of a percent of their daily revenue. In 2011, Tyson chicken had a deadly chlorine gas incident at a plant that injured nearly 200 people and could have killed them all. And after they were judged to be negligent by OSHA, and their negligence determined to be "serious," they were fined $7,000. In a year where Tyson posted $733 million of profits. That comes to 0.00095 percent. And then guess what they did? Contested it and only ended up paying out $2,500—$14 for each of the 173 workers who were hospitalized from the chlorine gas.

It's like your boss fining you a nickel for taking a two-hour lunch break. Welp, boss, here's a nickel, see you in a couple hours. I'm gonna take that two-hour lunch every day because this is a very good deal for me. Handing out these slap-on-the-wrist fines for violating safety standards or forcing people to work longer hours in unsafe conditions will never stop a company from breaking that law. Because it's a drop in the bucket. And while their workers are wearing adult diapers and literally shitting themselves while processing your dinner because they've got to make quotas,¶ Tyson continues to lobby aggressively to remove caps on the speed at which slaughter and evisceration lines can operate—because of course they are.

Sorry, I digress. We could write a whole book on all the ways corporations abuse and exploit humans for profit. Let's get back to abusing and exploiting the planet.

It should be so expensive to create waste as a company that they are fully incentivized to fix it. And then there's the other issue with this particular solution: most of the time, these costs are just passed on to the consumer in the form of price increases anyway. It may help from a materials standpoint, but it doesn't do much in terms of making companies pay for their wastefulness. Coca-Cola could just keep using virgin plastic for the rest of time, but that bottle of Coke will cost $5 instead of $1.

That's why these fines have to be absolutely outrageous. So outrageous as to make their business models unsustainable while paying them. And maybe put a hard cap on materials usage in the first place. Oh, and carbon emissions. You could make a Coke bottle that uses 90 percent less plastic, but if it has five times the carbon emissions . . . I'd rather just use more plastic, to be honest. I think? That's a tough one.

How about any company who can create a fully closed-loop system

---

¶   I think we're all aware Amazon is doing this too.

for their products where every material is fully re-used should receive massive tax breaks? Anyone wanna bring back the milk man??

Don't tax the people, tax the products.

Okay, we're starting to get a little more radical! Exciting! This is an interesting solution, and could be an answer to the problem presented in the last one. My podcast guru Nate Hagens is a proponent of some form of this in the future: stop taxing individuals. Tax the products, and tax the products according to their real material value and real harm to the environment. Oil may be cheap to get out of the ground right now, but its true value along with its environmental harm is far higher. Will a cell phone cost six thousand bucks? Maybe. But if you still want to pay for it, that's on you. Hopefully we won't have planned obsolescence, and the phone will actually last ten years. Products that don't harm the environment won't cost as much. For businesses, this would mean paying taxes at every step of the production process. Buying plastic for your packaging? Big tax. Buying cotton that takes an insane amount of water to grow? Big tax. Making your product out of environmentally friendly hemp instead? Smaller tax. Have a manufacturing process that involves shipping your unfinished product twelve times across the ocean to be assembled in twelve different factories? Big tax.

Shifting the tax base from labor (i.e. income tax) to consumption would incentivize the transition to more environmentally friendly products—because they would be the cheapest ones. We'd probably start seeing a lot more sustainable companies popping up as well.

The best part about this one is that every person gets to keep 100 percent of their income! You'll *only* pay taxes on the things that you buy. This has the added bonus of incentivizing people to buy less, which is something we desperately need. The downside is that consumption taxes are regressive, which means they affect poor people disproportionately. But maybe we could limit it to anything outside of basic human necessities like food, water, and shelter, at the very least. I'm open to ideas on how to counteract the regressiveness of it.

And let's unwind fossil fuel subsidies, obviously.

I honestly don't know how we are still doing this. And global fossil fuel subsidies are actually increasing! Everyone's scared about where they're getting their energy from, and so the most obvious answer is to make it even cheaper to dig into the earth and burst those shale pockets. But these subsidies are massive. And oil is already a thousand times cheaper than it should be. We don't need to make it any cheaper. We need to make it more expensive, so we stop using it so much. The United States provided $20.8 billion in direct fossil fuel subsidies in 2021, but that doesn't even include all the tax breaks they get. According to the IMF, the estimated cost of fossil fuel subsidies when you account for externalities (like environmental damage) was $5.9 trillion or 6.8 percent of global GDP. That's insane. Whenever you hear of a subsidy, it's important to realize that's something *you* are paying for. That's *your* money the government is spending. You are paying for oil companies to be able to do business for cheaper and make even more money . . . from you. Does that seem right to you?

Unwind half of military spending.

Okay, this one is less likely to happen when we're looking at a future that could be plagued by international resource wars amidst a geopolitical clusterfuck. But seriously, we spend wayyyy too much money on the military. And so much of that money could have been going to build a better nation. I feel like somebody could go in there and squeeze out at least a few hundred billion *per year.*

The National Defense Authorization Act (which determines our annual defense spending every year across the Department of Defense, Department of Energy national security programs, and the Department of State) gets released, amended, and passed separately from the overall budget. They have built it so it can't possibly get tied up with all those normal budgetary concerns like education and healthcare. And this massive—I mean truly massive—bill gets passed with broad support from both sides of the aisle every single year. It's the one thing everyone can agree on! Bipartisanship *does* exist! JK. The only reason this happens without fail is because they put allllll kinds of money in there. And everyone agrees that no one will vote against it. Consistently, this spending bill is approved for tens of billions of dollars *above* the budget request from the President. And not a single fiscal conservative can be heard saying "But where will we get the money for that?" The even crazier part? We just left a twenty-year war in Afghanistan—packed up all our shit and got the hell out—and the budget *still* got bigger. You would think that leaving a war that was costing $300 million a day* for twenty years would maybe like, save some money . . . but apparently not. Let's dig in a bit, shall we?

In 2022, the Biden administration requested $802 billion for national security. The House allocated $839 billion, or $37 billion more than they sought. That "National Defense Authorization Act" passed 329–101.

In 2021, the Presidential budget requested $744 billion. Congress instead allocated $768 billion—$24 billion more than requested, notably including $28 billion to build out our nuclear weapons program. That one passed 363–70 in the House and 89–10 in the Senate.

In 2020, Trump asked Congress for $750 billion. The final number landed at $778 billion, and it passed 86–8 and 377–48 in a Congress that was so divided they were literally trying to impeach the guy. And, just for reference, this 2020 number was already $100 billion more than the FY 2017 budget under the Obama administration, a measly $647 billion.

It just. keeps. getting. bigger.

So in the six years from 2017 to 2023, our military spending increased from $647 billion to $839 billion—$190 billion dollars, or a 29 percent increase. My only point in running though these absurd numbers is that I'm pretty sure we can find some fat to cut in there. I'm also pretty sure this is some secret money funnel into Congressional bank accounts. They can't even pretend to tell you where all this money is going.

The Department of Defense just failed their *fifth* audit in a row. They are the only government agency that has never passed a single audit

---

* I'm not even kidding. $300 million every single day. That's $109.5 billion per year.

since the law was introduced in 1990. They didn't even *try* until 2017. Only seven of the twenty-seven entities that make up the DoD received a clean bill of health . . . for an organization with $7,000,000,000,000 ($7 trillion) in assets. And even worse, the results of this audit were basically the same as last year. They've made no progress. And on top of *that*, the audit itself costs a billion dollars! Are you serious? Where the fuck is all this money going?

In 2015, the U.S. government had a total federal budget of $4.2 trillion. That's $4,200,000,000,000.[†]

But how much is a trillion, really? Imagine you had a billion dollars! Now imagine you had a thousand stacks of a billion dollars each. Now imagine you had four of those. And then you had another pile that was the entirety of Elon Musk's net worth (pre-Twitter) on the side just for a rainy-day fund. That's how much we spend every year. But really, even conceptualizing a billion is hard, so let's try another way. How long do you think a million seconds is? It's about 12 days. And how long do you think a billion seconds is? It's 32 years. Okay . . . now guess how long a trillion seconds is.

It's 31,688 years. How many years did you guess? 500? A thousand? The extinction of the woolly mammoth only happened 13,000 years ago . . . or less than half a trillion seconds. Whatever you imagine when you're imagining a trillion dollars—you're doing it wrong. A trillion is, for our little pea brains, quite literally unimaginable.

Personally, I agree with Obama's defense secretary, Robert Gates: "If the Department of Defense can't figure out a way to defend the United States on half a trillion dollars a year, then our problems are much bigger than anything that can be cured by buying a few more ships and planes." Amen, brother.

One more fun fact while we're at it: the US military was responsible for 51 million metric tons of CO_2 in 2021, more than most countries. So mayyyyybe there are a couple good reasons to take 'em down a peg or two.

---

† I'm continuing to write out all the zeros in trillion until I can wrap my head around it. The only effect it's having is making millions look like barely a number at all. Yikes.

2015 Federal Income Tax Receipt

April 15, 2016

Health	$3,728.92
Includes $1,702.24 for Medicaid	
Includes $44.94 for Children's Health Insurance Program	
Pentagon & Military	$3,299.13
Includes $706.70 for Military Personnel	
Includes $90.97 for Nuclear Weapons	
Interest on Debt	$1,776.06
Unemployment and Labor	$1,040.93
Includes $77.59 for Temporary Assistance for Needy Families	
Includes $29.45 for Job Training and Employment Programs	
Veterans Benefits	$771.26
Includes $339.33 for Payments for disability, death, etc.	
Includes $309.21 for Veterans Health Administration	
Food and Agriculture	$598.74
Includes $370.57 for SNAP (food stamps)	
Includes $35.40 for Federal Crop Insurance	
Education	$461.59
Includes $153.74 for Pell Grants, Work Study, and other Student Aid	
Includes $109.03 for Elementary and Secondary Education	
Government	$377.50
Includes $61.34 for U.S. Customs and Border Protection	
Includes $34.29 for Federal Prison System	
Housing and Community	$250.03
Includes $48.55 for Head Start and Related	
Includes $9.24 for Homeless Assistance Grants	
Energy and Environment	$207.68
Includes $42.07 for Environmental Protection Agency	
Includes $9.14 for Energy efficiency and renewable energy	
International Affairs	$194.29
Includes $38.63 for Diplomatic and consular programs	
Includes $33.68 for Global Health Programs	
Transportation	$150.68
Includes $25.08 for Transportation Security Administration (TSA)	
Includes $5.50 for Federal Aviation Administration	
Science	$143.20
Includes $86.11 for NASA	
Includes $33.08 for National Science Foundation	
TOTAL	$13,000

See and Share the Average Federal Income Tax Receipt for Your State

Now let's look at it another way. Did you know that in Australia, whenever you get your paycheck, it tells you exactly how much you paid for all of the government services they provide? It's great! You wanna know why we don't do that in America? I'll give you one guess.

Now let's travel back in time to 2015 and take a look at the taxes of the average American. Why 2015? Well, partially because the last couple years of tax revenues and spending got pretty crazy with the pandemic, but mostly because CNN Money[88] already did all this math for me. According to their fun infographic, the average American paid $13,000 in federal taxes in 2015. That includes federal income tax, social security, and Medicare, but not taxes you pay to your state such as property taxes and car taxes and additional income taxes.

If the average American paid $13,000, you can see where it went in the chart on the previous page. And here's a pie chart of the same information because pie charts are helpful.[‡]

Average federal income tax spending, 2015

Based on average 2015 individual federal tax bill of $13,000

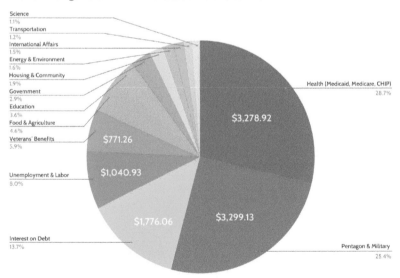

Science
1.1%
Transportation
1.2%
International Affairs
1.5%
Energy & Environment
1.6%
Housing & Community
1.9%
Government
2.9%
Education
3.6%
Food & Agriculture
4.6%
Veterans' Benefits
5.9%

Health (Medicaid, Medicare, CHIP)
28.7%

$3,278.92

$771.26

Unemployment & Labor
8.0%

$1,040.93

$1,776.06

$3,299.13

Interest on Debt
13.7%

Pentagon & Military
25.4%

Source data: National Priorities Project interactive tax calculator

‡   You can use this handy calculator here to see the same breakdown for your 2022 taxes: https://www.nationalpriorities.org/interactive-data/taxday/average/2022/receipt/

Do you think if people knew they were only paying $29.45 for job training and employment programs, that they might be willing to throw an extra $10 at it to help retrain coal mine workers to transition into new industries? What about the fact that you're only paying $9.24 for homeless assistance grants? $9.14 for renewable energy? Even only $370.57 goes to the oft-villainized SNAP program, which provides food security for the entire country. If we halved military spending in this example, you'd have an extra $1,640.57 of *your* tax dollars to allocate to all sorts of different programs ... where do you think you'd put it?

And halving military spending would only put us back to around 2003 levels! Totally doable! And still more than the next top three countries combined in 2021 (China, India, and the UK). If you were around in 2001, I'm sure you're well aware we were putting plenty of money towards the military in the years immediately following 9/11.

Want some more money? Raise the social security income cap.

If unwinding the military doesn't sound like an easy way to get more money to help more people, I've got an even easier one! For some asinine reason, people only pay social security tax up to $127,000 in income. I've got a real easy way to generate a whole lot of extra money for the government. Start making people who make more than $127,000 continue to pay social security tax. And guess who this will harm? Only people already making more than $127,000. I continue to be confused how we're not already doing this.

Universal basic income.

Why do we need all this extra tax money? Well, hear me out. What if instead of attempting to incentivize private companies to make decisions that are better for humans, we just started paying for all the things that keep people happy and healthy? Housing, food, healthcare, water, energy, local transportation, community, and education? It's not that far from where most social-democratic nations are now. It's not very far from where we are now in theory, minus the whole healthcare debacle, and gutting our welfare programs, but we pay for all of these things to some degree in the United States. It's just means-tested, which means you have to jump through a whole bunch of hoops to qualify for the assistance. And our government keeps making those hoops harder and harder to jump through. And we have an entire bloated industry dedicated to stopping you from making the jump. And if you do manage to get on the assistance, people will call you a lazy piece of shit and then mumble something about bootstraps. And even worse, you'll call yourself a lazy piece of shit because this narrative that our personal worth is tied to our ability to make money is so deeply ingrained in our national ethos.

Anyway, if you really hate handouts to those lazy welfare queens ... I've got a solution: Universal Basic Income. If you're not familiar with this, it's exactly what it sounds like. Instead of allocating a little money to food stamps, and a little money to housing vouchers, you just give everybody money. Basically the smallest amount of money you would need to live on. Say, $12,000 a year. And everybody in the country over a certain age gets this regardless of whether they work or don't work. Some UBI proposals have an upper-limit cap, so people making over $100,000 or $150,000 a year would be ineligible. Which I'm okay with as well.

One of the benefits of this is it drastically cuts government bloat. No more need to spend billions of dollars on the offices that make you file dozens of different forms each year to qualify for your assistance. One of the main arguments against it is that it will make people lazy, and they'll just mooch off the system, and we NEED people to WORK.

You're not wrong. Some people will be lazy and just live on $12,000 a year. But the wild majority of people do not want to live on that little. Do you want to live on that little? For most people, it would act as more of a safety net to try something new, to quit that job you hate, or start working part-time so you can spend more time with your kid, or write that book you always wanted to write. Or just because working part-time and relaxing the rest of the time is a totally valid decision. And for other people they'd probably just put it in their savings with the rest of their money or take an extra fancy vacation. It doesn't really matter.

While we're at it, let's make care work paid labor!

Oouu man, we're really starting to push the boundaries here, and I like it! One of the biggest oversights of the capitalist economic model (other than that whole 'not accounting for energy' thing), is not accounting for care work like raising babies as a valuable economic input.

It's mind-boggling to me that we somehow neglected to realize that if people don't stay home to have babies—who then become adults who get jobs and go to work—there IS no economy. There is one thing and one thing only that keeps an economy running in the long run: more fucking humans.

So let's make care work paid work. Let's make it so that people who have to stay home taking care of their kids or a disabled sibling or their aging parents are compensated for that work. It's valuable to the economy, and since the government loves protecting the economy at all costs, make them pay for it!

Think about it like this: If a man hires a housekeeper to clean his house and cook his food, he has to pay him (or her). But if he decides to start shacking up with that housekeeper, and then they get married, and then she just stays home and does the exact same work except without a paycheck . . . where did that money go? One day, the work she was doing went from having a countable monetary value to . . . not. Care work is necessary to have a functioning economy—to create the humans economies need in order to self-sustain—but until we start valuing that, it's going to remain, well, unvalued.

And it's definitely not helping that raising a child in America now costs a quarter of a million dollars. All these countries with low birth rates are running ad campaigns encouraging procreation when all they have to do is make it actually affordable to have a baby.

Now ban living off of interest.

Speaking of valuable unpaid jobs, now let's talk about unvaluable overpaid jobs. This one is a lot harder to ban. I can't exactly stop you from living off the interest that the $10 million you have sitting in the bank generates every year. In case you're wondering, at today's current interest rates, you could have ten million sitting in a *checking account* and you'd have an income of . . . $400,000 per year. For doing nothing. And if you don't spend the whole 400 grand, it will start compounding, and next year you'll earn even more! For continuing to do absolutely nothing. As OG ecological economist Herman Daly once said, "What is obviously impossible for the community—for everyone to live on interest—should also be forbidden to individuals, as a principle of fairness."

So you live off interest from loans you made to other people or money you left sitting in the bank or in the market. But if everyone did that? Well, everyone can't do it. Because then who would you be loaning the money to? It's creating a class of people who produce no actual value in the world, and ironically, provides them with the most wealth while they do it. Pretty screwed up, huh?

So we have to severely restrict income that isn't dependent on labor or actual work, but is just derived from renting out land or buildings or checking accounts. Once again, btdubs, this is NOT a new idea.

Physiocratic economist François Quesnay—back in the 1750s— argued that taxes on productive classes, like farmers, should be much lower than non-productive classes, like landlords and capitalists (owners of the businesses that make the monies). Because obviously. One of them is producing something valuable and the other two are . . . not.

But how would it work? Good question. Since banning it outright isn't really feasible, just tax the shit out of it. That might be my new motto, actually. Tax the ever-loving shit out of capital gains and rental income. Tax the ever-loving shit out of anything that doesn't provide actual value to our society. Right now, short-term capital gains from when you sell a stock in less than a year, or when your checking account earns $400,000 in interest, are taxed at the same rate as your income. The higher your total income, in theory, the higher rate you'll pay. But long-term capital gains (when you sell stock you've owned for more than a year) are taxed wayyyyyy lower and are capped at 20 percent for even the richest motherfuckers in the country, even though the top marginal rate for income tax is 37 percent. That's right: we tax money "earned" when rich people just sit on their money at almost half the rate we tax people who actually worked for it. And the estate tax for people inheriting large sums of money? Well, it's been gutted to make sure they can keep all that money in the family.

Considering the wealthiest ten percent of Americans own 89 percent of the stock market,[89] I'm not too worried about who this is going to hurt.

Oh! And there's another fun one known as the "carried interest loophole." This one means hedge fund managers—who definitely without question get paid regular salaries—get to treat those salaries as long-term capital gains instead of income, and thus only pay half the taxes that a normal person would. It's estimated closing this loophole could generate between $1.4 billion and $18 billion annually. You and I are literally giving them this money. Once again, whenever you hear about a "tax break," think about it like money going out of your pocket and into theirs. Poor people are subsidizing the salaries of the ultrawealthy. How absurd is that?

Forgive all debt. Then abolish it.

If we're going to ban living off interest . . . what if we just banned debt altogether??? Okay, I admit, this one is a little crazy. And I have no idea how it would work. But since a society built on debt isn't even theoretically feasible, much less actually, practically feasible, we've gotta figure out something here. So let's wipe out the personal and business debt of everyone in America. We wouldn't even be the first society to do it.

In 2400 BCE the Sumerian King Enmetena declared a debt amnesty. In 1792 BCE, Hammurabi declared a debt amnesty. In ancient Israel they declared a debt jubilee every fifty years. It was the only way to rebalance their societies that were already suffering from extreme inequality and unrest.

The thing about debt is that it endlessly fuels inequality. The rich get richer and richer by design. There's no escaping it. People who are already wealthy do not take the interest they earn and go spend it buying milk and eggs. They're already buying all the things they want. So they hoard it. And then it earns more interest.

What would the consequences of debt jubilee look like? Well, a bunch of financial companies would lose a bunch of money. Some would collapse. But I think we'd be fine if we did it right.

Now, I'm fully aware that debt provides opportunity for people who don't have big piles of cash sitting around to do things that improve their lives. But perhaps there are some things we could do to limit corporate debt while still allowing personal debt to go to college or start a business?

Guess what, there is! I just googled it, and now I am a little bit smarter. Thank you, internet! It's called a positive money system.

We get rid of debt-based currency altogether and invent a whole new system. In the new system, rather than banks "lending money into existence," governments create money by "spending it into existence." Banks can't create negative pigs. And all the money that exists in the real world has gone to real actual people doing real things. Not just going straight into financial speculation that benefits the ultra-wealthy. Who gets to decide how much money we create? I don't know, some kind of government body. But at least it will be a body specifically tasked with upholding the public good and not, you know, perversely incentivized by corporate profits and private jets.

In this system, banks can still lend money to people to buy a house or go to college, but they have to have 100 percent of the money they lend in reserve. No more negative pigs. This isn't a new idea, by the way. Apparently it's been around since the 1930s when reckless lending led to the Great Depression, and was brought up again in 2012 in the wake of the Great Recession. A recession triggered once again by . . . reckless lending. Not only could it solve our banking issues, but it's also a pretty solid step toward curing our addiction to endless growth. We just have to forgive all the existing debt first to get started, nbd, right? According to Jason Hickel with *The Guardian*:

> The idea has its enemies, of course. If we shift to a positive money system, big banks will no longer have the power to literally make money out of nothing and the rich will no longer reap millions from asset bubbles. Unsurprisingly, neither of these groups would

be pleased by this prospect. But if we want to build a fairer, more ecologically sound economy, that's a battle that we can't be afraid to fight. [90]

I know it's far more likely that we go the completely opposite direction and continue to make ever-more-complex financial levers to kick the can of impending global financial collapse just a little further down the road, but hey, a girl can dream!

Decouple education spending from property tax.

What does this one have to do with the economy or the environment? Nothing. It's just better for humans. Kids perform better in more diverse schools—both rich kids and poor kids. They also become more culturally understanding and a wealth of other things. There has been study after study showing that diversifying schools is better for everyone. So why haven't we done it? Racism.

Basing school funding on property tax was just a lever to create areas of urban blight in the 1950s where the black schools got worse and worse, they moved them into neighborhoods with little to no community resources, next to the housing projects that were just being developed, with no parks or playgrounds or libraries, further away from any white kids who might have gone to the previously integrated school.

We did this on purpose. Systematically.

One of my favorite/least favorite stories is that of busing in North Carolina. North Carolina had a busing program that drove kids from the poor neighborhoods with the shitty schools into the better neighborhoods with the better schools. It was a part of active desegregation in the wake of school integration mandated by *Brown v. Board of Education*. Charlotte schools even got their own Supreme Court ruling mandating busing in the 1970s after the city and county had cleverly avoided integrating their schools for like twenty years.[§] The thing is, not only did the poor kids do better, so did the rich ones. The city was hailed as a wild success in school integration! The poor kids' academic

---

§   They were reallllly hoping that Supreme Court case would go the other way.

performance improved, the affluent little white kids saw their test scores improve, and everybody was happy for like a good twenty years until . . . they canceled the program. Why? I'll give you one guess. It starts with an 'R' and rhymes with shmay-sizm. White parents complained to the school board and got the busing program canceled.

The whole thing went to federal court, where Judge Robert Potter racistly argued that since Charlotte had "eliminated, to the extent practicable, the vestiges of past discrimination in the traditional areas of school operations"[91] that they weren't federally required to keep the busing program in place. Another popular re-segregation argument of the 21st century (there have been a LOT of court cases about school integration) is that since schools are not allowed to use race as a factor in assigning school districts (an *anti*-segregation ruling), they also couldn't use race in the opposite direction—to make it better.

Well played racists, well played.

So the busing was ended and students went back to attending schools in their own neighborhoods. But because the majority of the neighborhoods in Charlotte had remained racially segregated the whole time the busing program was in place—and black neighborhoods are just about always economically disadvantaged compared to their white counterparts—the school achievement gap immediately grew larger. Because schools get funded by property taxes. And aside from less diverse schools just being bad for kids, this new system also contributed to the concentration of social, political, and economic resources in a handful of white neighborhoods (shocker). So pretty quickly in the wake of the 1999 un-busing decision, the Charlotte-Mecklenburg school system went from being a model of racial integration to one of the most segregated school districts in the country. And similar decisions took place all across the country in the early 21st century. Cool.

So anyway, yeah. Why don't all the school districts in a state just share all the money for schools and then all the schools have the same quality of education, and we don't further increase social inequality by keeping the poor kids poor and dumb? Oh right, I guess that was the whole point.

Fun fact! In Norway, it's illegal to charge tuition for schools, so all the rich kids have to go to school with all the poor kids, so the rich parents are incentivized to make the schools really good, and everybody wins! Hooray!

Revamp countrywide zoning laws on affordable housing and multifamily homes.

Speaking of social inequality and wealth inequality and systemic racism . . . here's another fun one that has absolutely zero to do with the environment. Let's revamp zoning laws across the country! Does that sound extremely unsexy to you? Well, just you wait. It should come as no surprise that the United States has a rich and vibrant history of racial discrimination in housing. Those discriminatory policies (that were federally mandated to keep black people poor) led directly to the wealth gaps we see between black and white today. This wasn't some, "Oh you went to a poor school so you got a worse education so you never got a good job" argument. This was: "I am going to stop you from owning a home by physically making it illegal for people to build one for you or sell one to you—while giving major benefits and tax breaks and down-payment assistance to white homeowners—codifying the major driver of white wealth today" kind of a thing. Black people don't live in shitty ghettos because they're lazy. They live in shitty ghettos because we created the ghettos and forced them to move there. We zoned neighborhoods so they could only have big, single-family houses that only white people could afford. And we made it so the only places black people could buy houses were the same places zoned for strip clubs and industrial waste. For good measure, we also made it illegal to sell your nice suburban house to a black person even if you wanted to and even if they somehow managed to scrounge up enough money to do it.

But now, even today, multi-family units cannot be built in these bougie neighborhoods. Even though we have a massive affordable housing shortage, and multi-family units don't increase crime, or get all the kids to start smoking crack, or whatever other arguments people have against them. And once again, Norway has managed to sprinkle cheaper, multi-family units among the larger single-family homes in nice neighborhoods without building massive high rises or turning their suburbs into blight-ridden crack dens.

Just like the education funding issues, the more we DE-segregate our country, the better it is for everyone. This isn't a book about systemic racism.* But lowering inequality—both racially based and the good ol' fashioned billionaire-versus-everyman kind—is good for society. And this book is about things that are good for society.

And while we're talking about zoning, it actually also has a lot to do with the environment. Maybe we can help black people *and* the environment. Current zoning laws don't just keep poorer people out of fancier school districts. They also keep certain areas residential and other areas commercial. Which means you have to get in your car to drive anywhere. Usually to a heinous strip mall with a Walmart and a Chick-fil-a and a T.J. Maxx and Panera Bread. Or insert your local big-box store and fast-food chains here. But what if every American neighborhood was required to have a small market, a restaurant, and a couple shops within walking distance? What if they were purpose-built to keep people out of their cars instead of the other way around? We're not going to raze every suburban subdevelopment that's been built since 1950. And adding public transport into shopping centers doesn't keep people out of their cars. But making the communities people live in more livable and walkable does.

And maybe we could also tear down all the highways we built specifically to decimate black urban communities and replace them with all this wonderfully robust public transportation we need while we're at it?

Okay, I think you get the idea.

---

* If you want to read a great one on the topic, check out Richard Rothstein's *The Color of Law.*

The Cultural Shift We Need

Where are Maynard Keynes' grandkids?

As far as I have heard or observed, the principal object
is not that mankind may be well and honestly clad, but,
unquestionably, that corporations may be enriched.
In the long run men hit only what they aim at.

—Henry David Thoreau
Walden

If we've learned anything from that last section (and the one before
that), it's that corporations operating on a purely profit-motivated basis
in a corrupt system where they get to design the rules of the game is
clearly bad for the individual. It's bad for me, it's bad for you, and it's
certainly bad for the environment.

But even if you still don't think we're gonna run out of oil or you
want to ignore all the hurricanes and floods and droughts and fires, you
can't deny the fact that society feels pretty sad and fragile right now. You
can't deny the tribalism and the depression and the alienation and so
many other messed up things happening around us. Even if you think
we'll make it through just keeping on keeping on doing whatever we're
doing, you can't deny that late-stage, endless growth capitalism is kind
of making us . . . miserable.

Let's say we perfect clean hydrogen fuel *and* commercialize fusion or deep geothermal *and* find a way to capture all the existing CO_2 *and* stop climate change in its tracks *and* save the ecosystem and the bees and the coral reefs *and* find an asteroid with every rare metal we need *and* solve political polarization, and we can finally go right back to ordering Uber Eats from our La-Z-Boy recliners while binge-watching Netflix and simultaneously scrolling our phones . . . do we even want that?

Is endless growth capitalism making us happier?

Is it making *you* any happier?

We've talked plenty about perverse incentives as they relate to corporations making decisions that benefit only corporations. But we haven't yet talked about the individual. We haven't talked about *you*.

We know because of the wealth-accumulation obligation, individuals, businesses, and nations are incentivized to do everything in their power to accumulate more wealth. The problem is, in our outdated neoliberal economic models, the source of all that wealth is human labor. It's us. Human labor is the biggest cost input into their cherished circular-flow. And if you're a business owner, anything that costs money in your model should be minimized at all costs. Hence why we replace those radish-picking humans with a much cheaper machine.

Recently, the *New York Times*[92] asked a bunch of twelve- to four-teen-year-olds what makes them nervous or scared about getting older, growing up. Wanna know what they said?

- **Jillian, 14:** "I'm scared about finances. You have to pay for college and then a house. There's just a lot of expenses."

- **Wynter, 14**: "I'm scared of the possibility of losing my job and not having a stable income to get a house or apartment to live in."

- **Andrew, 11:** "I'm scared of getting too caught up in work and separating from people I care about."

- **Matthew, 13**: "I'm scared of failing in whatever my job is and not having money."

- **Jonah, 12:** "I'm not really scared of anything."

I mean, let's all be more like Jonah, and maybe Andrew, but I don't think twelve-year-olds should be having these thoughts, do you?

The thing is, most of us know that our bosses don't give a shit about us. Your kids already know that. We know that the second it's cheaper for a machine to do our jobs, we're getting the axe. We're only employed because for the time being, our labor is cheaper than a machine or a machine just can't do it yet. If we've learned anything in the years since the automation boom and the release of ChatGPT, it's that this is only becoming more true for more jobs. Even the fucking twelve-year-olds are worried about it.

The core tenet of the system is designed to exploit humans as much as possible, or ideally—take us out of the equation altogether.

We are a cost input to be minimized, a line item in a spreadsheet. And if they manage to find a machine to do your job, and you can't find another company willing to exploit you, you're shit-out-of-luck. Maybe you end up homeless or maybe you starve to death if you live in a country without social safety nets. Or maybe you just spend the rest of your life struggling to make ends meet, choosing between feeding your kids and turning on the heat, and hating yourself for being such a failure. Because capitalism has also told you it's your fault if you don't succeed in a system designed to use us up and throw us out. At least the serfs always had a house.

If we're looking for a better system—maybe a society where people are actually *happy*—we might want to consider a model where humans aren't looked at as an ideally expendable piece of a money-making machine that could toss them to the curb at any moment.

Let's talk about making aglets. What's an aglet? It's the little piece of plastic that goes at the end of your shoelace. The aglet is a real thing, but this example is fictional. So in the whole world, let's say there are two aglet factories that each employ fifty people. There are a hundred people working eight hours a day to make all the aglets the world needs.

At the end of the day, everybody goes home, and no one is left wanting for aglets. But one day, somebody makes an invention that cuts aglet production time in half! Now it only takes four hours a day to make all the aglets. Hooray! Everyone gets afternoons off! Except of course that's not how it happens. The people continue working eight hours a day at both factories, they produce more aglets than anyone could ever want, the price of aglets collapses, one of the aglet factories goes bankrupt, and half of the people who used to make aglets are now broke and unemployed. The messed-up part is there is still just as much free time being divided between these workers, and the same number of aglets are being made and sold—it's just that half the people are overworked while the other half are idle and destitute. The ones who got fired, rather than getting to enjoy their leisure, are forced into a state of miserable loafing and crises of self-worth, while the other guys don't get to enjoy anything at all because they have to slave day and night at the aglet factory. As Bertrand Russell said when he first penned a version of this example, "Can anything more insane be imagined?"

The even worse part is that lady who kept her job at the aglet factory is still struggling. Endless growth means inflation makes her cost of living higher and higher every year. And her boss never gives her a raise because minimizing the human labor cost is her boss's only objective: to give her the smallest amount of money possible to keep her doing her job. Any raises ever given to anyone in capitalism are only to keep them from quitting so they can continue funneling profit and value to the owner of the business.

The more you think about the dynamics at play within capitalism, the more back-asswards it feels.

The argument goes that capitalism drives innovation that means our lives have improved. Without capitalism no one would invent anything because what's the point if you can't get rich? And letting the markets do what they do is the only way to really breed that innovation. Without capitalism, we'd all be sitting at home on our asses just melting into the couch because humans have no inherent motivation to do anything. JK. Except that's the story that capitalism has decided to tell. But other than the wonderful, incredible innovation that's brought us smartphones and

DiGiorno Rising Crust Supreme, it seems like a system that's actually designed to benefit the *fewest* people rather than the most—and this kicks in the MOMENT those businesses grow beyond the size of the communities they serve, the moment they stop considering the communities they rely on and the lives they affect. And the bigger they get, the worse it gets because that's how the system is fucking designed. It's not a bug, it's a feature.

Haven't you ever played Monopoly? It's literally how it goes every single time. Once one person gets a couple key properties, they continue to amass more and more while the other players start going bankrupt. One stroke of luck and the playing field is forever unleveled. Suddenly, rent's increased by four-fold, and then your brother throws the whole boardgame across the room when he lands on Boardwalk with two hotels. Once one person gets all the money, the game ends because that's how it always ends. The game is called Monopoly for a real good reason. Because all capitalist systems tend toward monopoly. The lady who invented the game, Lizzie Magie, did so to prove a point about the enrichment of property owners and impoverishment of tenants. It's a game based on capitalism that works exactly the way capitalism is built to work.

All of the massive increases in productivity over the last hundred years, kicked into overdrive since the dawn of the internet, have gone to those eight white dudes who own half the wealth in the entire world. You're still poor, I'm still poor,[†] and it's only going to get worse. According to the authors of *Tightrope*,[93] if we had the same income distribution we had in 1979, the bottom 80 percent would have $1 trillion more, and the top 1 percent would have a trillion less. That's a thousand billion dollars they wouldn't have. And it would mean the average working family would have seen an almost 25 percent increase in income.

Ironically, early economists truly believed that the system was going to make things better and make people happier. They believed all this innovation was going to allow us to work LESS. Please refer to the aglet example to see how that definitely hasn't happened at all.

---

† On a global scale, of course, with a household income around $90,000, I'm actually one of the richest people in the world. You probably are too. And yet somehow we're *still* poor in America.

Even Maynard Keynes believed his grandchildren would only work fifteen hours a week. In his 1930 essay entitled, "Economic Possibilities for Our Grandchildren" Keynes predicted that the increases in technology and machinery and productivity would mean we would all have the luxury of working two days a week and devoting five to leisure. Or you can work five, three-hour days if you want . . . whatever. Just like Adam Smith saw capitalism as a force for social good in the 18th century, there was this idea back in the 19th century that once we had enough stuff, we'd all be like, "Oh cool, now I can just go lie in my hammock and write poetry." All those fossil fuel slaves are going to do our jobs for us! They're gonna save us so much time! All of the mind-blowing conveniences of the Incredible Everything Machine like spinning jennies and fax machines would mean more free time, which means more time to enjoy our lives, enjoy our loved ones, travel, relax, create art, enjoy life!

John Stuart Mill, in his 1848 essay arguing for the end of endless growth, saw both the potential for capitalism to provide that freedom and social progress and argued that its most "legitimate effect" was abridging labor. Amen, brother.

> It is scarcely necessary to remark that a stationary condition of capital and population implies no stationary state of human improvement. There would be as much scope as ever for all kinds of mental culture, and moral and social progress; as much room for improving the Art of Living, and much more likelihood of its being improved when minds ceased to be engrossed by the art of getting on.
>
> Even the industrial arts might be as earnestly and as successfully cultivated, with this sole difference: that instead of serving no purpose but the increase of wealth, industrial improvements would produce their legitimate effect, that of abridging labor.[94]

Once we have all this free time and social progress, our lives are going to get even better! Material prosperity from capitalism will give everyone "time prosperity." We'll be rich in money and rich in time and the forty to eighty hours a week most of us spend working will be replaced by the search for the good life. We'll spend our time seeking pleasure and meaning. Or maybe many of us will keep working, but it will be pleasurable and useful. The people who always wished they had time to write a book could finally write that book.

Keynes believed when our basic needs were so comfortably met, our sense of what is truly important would change organically, and one day we'd just magically recognize that "avarice is a vice, that the exaction of usury is a misdemeanour,‡ and the love of money is detestable."

But of course, that's not really how it's played out. And John Stuart Mill picked up on that as well. Here's the next paragraph from that same essay:

> Hitherto it is questionable if all the mechanical inventions yet made have lightened the day's toil of any human being. They have enabled a greater population to live the same life of drudgery and imprisonment, and an increased number of manufacturers and others to make fortunes. They have increased the comforts of the middle classes. But they have not yet begun to effect those great changes in human destiny which it is in their nature and in their futurity to accomplish.

We haven't even figured out how to get everyone's "basic needs so comfortably met" in the richest country in the world at the fucking zenith of capitalism. Nearly 38 million people are still living in poverty in the United States. That's almost 12 percent of the population.[95] How's capitalism working for them? Oh right, those 38 million people are just lazy. I forgot.

For the rest of the bottom 80 percent of the population, we're just struggling to get by in this life of "drudgery and imprisonment," paying off our student loans while spending half of our shitty paychecks on rent because from 1985 to 2020, the national median rent price rose

---

‡ Usury is the practice of lending money at obscene interest rates. I probably don't have to tell you, this is only getting worse.

149 percent, while overall income grew just 35 percent.[96] And that 35 percent growth *includes* rich people whose incomes grew a whole lot more than the rest of ours. Guess which socioeconomic class saw their incomes grow the least? Ding ding ding! The poor got poorer!

So, no, most of us can't really think about writing poetry in a hammock just yet. But up till now, even for those people who *could* afford to, almost nobody stopped working. Why not?

Because capitalism told us the story that even making pretty good money at the aglet factory isn't enough. You should STILL want a better job. Because we've all been told that working in a factory is for losers, and the only way to really be successful is to own the factory.

Here's Peter Whybrow from his recent book, *American Mania: When More Is Not Enough:*

> As measures of personal achievement alternative to money have declined in social significance, the options for standing aside from the "mad race to riches" have diminished. In America it is the cultural imperative that one strives to maximize one's financial success. [...] Prosperity's pursuit demands a single-minded, lifelong commitment. In the promised land, enough is never enough. [97]

How exactly did this come about? Because the embedded growth obligation of capitalism means companies *have* to grow. And the only way for them to keep growing is to get you to keep buying more things. As an aglet-making or radish-picking human, you're just trying to get whatever job you can so you can eat. We used to hunt mammoths to eat, then we farmed to eat, now we sit at our desks and go to pointless meetings to eat. The work isn't meant to be our purpose; it's meant to be a means to an end. But we've lost the plot.

Instead, we've been conditioned to want to achieve ever higher levels of wealth. They invent demand for needless products under the false pretense that things provide joy. And since there are always more things to buy, there is always a higher level of happiness you can achieve. Sure, you have money now, but if you just earn a little more, then you'll be really happy. It's very unlikely you ever end up owning the aglet factory, but the belief that you might keeps you toiling away.

And then even if you do end up owning the aglet factory, it's *still* not enough. You'll only really be successful if you own two factories. Or ten. There are literally billionaires who still want more money. Because money provides status and recognition. Because it's the only way we measure achievement in our country. And because our culture values money over happiness (or perhaps conflates the two, believing money IS happiness) we all just keep grinding waiting for that time when we're finally, magically happy.

So you think, maybe I must be doing it wrong. Maybe if I just buy the next thing, then I'll get to this elusive happy. But it's a trick. It's a mirage in the distance.

Here in the United States, at the heart of capitalism's rampant success, our ever-higher economic output isn't making us any fucking happier. The market worked for a minute there, back when you still spent two weeks a year making candles, but we've flown far past the expiration date.

The problem is that there is a limit. Moving from a mud hut with no heat, electricity, or running water into a warm house with access to education will definitely make your life better. Getting a job that pays enough that you're not constantly worrying about making rent and affording groceries and getting your car fixed when it breaks down will definitely make your life better. The stress of poverty literally kills you. But going from a job that pays $70,000 a year to $80,000 will not. Upgrading from a Hyundai to a Lexus or 2,000 square feet to 5,000 will not. We've just all been told it will. Continuing to work, continuing to grow our economy and buy more shit, just doesn't improve our quality of life anymore. The things that make it better are leisure time, family time, free time. Freeing ourselves from the pressure to constantly buy newer, better things makes us happier. We've got it backwards. The accumulation of stuff is not even remotely synonymous with the general happiness of a community.

The crazy part is the people running the show know the whole system falls apart if we all decide working fifteen hours a week is enough, and we'd rather just spend time with our families. Look at how much economists and corporations freaked out during "The Great Resignation." *What do you mean you don't want to work???* The economy was "struggling"* as employers couldn't find people willing to work for $7.25 an hour because people finally decided that making less money to stay home with their kids was maybe . . . better?

John Stuart Mill had similar thoughts about what the Industrial Revolution had done for mankind, from that same banger of an essay:

> I confess I am not charmed with the ideal of life held out by those who think that the normal state of human beings is that of struggling to get on; that the trampling, crushing, elbowing, and treading on each other's heels which form the existing type of social life, are the most desirable lot of human kind, or anything but the disagreeable symptoms of one of the phases of industrial progress.

And this whole time we're devoting our nasty, short, and brutish lives to accumulating more money and more stuff, we're getting busier and busier and less and less happy. Swedish economist Staffan Linder calls this the "harried leisure class." We're obsessed with optimizing not only our work time, but our free time. We try to maximize everything because we were raised in a society that values maximization. We even try to be efficient in our sleep! Biohacking and subconscious learning and keto-hacking and atomic habits and the Pomodoro technique and all kinds of hacking so we can unlock some next level of productivity as if we're not already working hard enough!

Throw in a Puritan-Protestant work ethic inspired by the biblical notion that "idle hands are the devil's playthings" and people believe not working your ass off for someone else to make a profit is a literal sin. Which is a pretty useful sentiment for the CEOs and world leaders. You know, because working all day really gets in the way of people figuring out how messed up it is that they're working all day and still barely hanging on . . . or how we still don't have fucking healthcare.

---

* Read: big business had to start paying a living wage.

And if we had more free time, we might, I don't know, start fighting for a better system.

And the things that get sacrificed at the altar of capitalism and productivity are time-intensive activities with no financial gain: friendships, childcare, relaxation, meditation. Consumption and culture have become another piece of the capitalist machine. Our culture IS consumption. They're not just exploiting our time and bodies while we're at work—they're exploiting what we do with the little time and money we have to spare.

All this time we have devoted to getting richer—which was originally meant to give us the ability to do the things we want—has crowded out the intimacy and connection on which we thrive. It's replaced the one thing that actually works at making us happier. It's the means that have become the end as most of us have forgotten why we were supposed to want the money in the first place: to be comfortable, to work less, and to enjoy the things we want to spend time on.

Adam Smith's other most famous work, *A Theory of Moral Sentiments*, tells a shockingly relatable story about the mindless pursuit of wealth. It's long, but it's worth reading to show you how not-at-all-new these ideas are. This was written in 1759:

> The poor man's son, whom heaven in its anger has visited with ambition, when he begins to look around him, admires the condition of the rich. He finds the cottage of his father too small for his accommodation, and fancies he should be lodged more at his ease in a palace. [. . .]

> To obtain the conveniences which these afford, he submits in the first year, nay in the first month of his application, to more fatigue of body and more uneasiness of mind than he could have suffered through the whole of his life from the want of them. He studies to distinguish himself in some laborious profession. With the most unrelenting industry he labours night and day to acquire talents superior to all his competitors. He endeavours next to bring those talents into public view, and with equal assiduity solicits every opportunity of employment. For this purpose he makes his court

to all mankind; he serves those whom he hates, and is obsequious to those whom he despises. Through the whole of his life he pursues the idea of a certain artificial and elegant repose which he may never arrive at, for which he sacrifices a real tranquility that is at all times in his power, and which, if in the extremity of old age he should at last attain to it, he will find to be in no respect preferable to that humble security and contentment which he had abandoned for it. [. . .]

It is then, in the last dregs of life, his body wasted with toil and diseases, his mind galled and ruffled by the memory of a thousand injuries and disappointments which he imagines he has met with from the injustice of his enemies, or from the perfidy and ingratitude of his friends, that he begins at last to find that wealth and greatness are mere trinkets of frivolous utility. [. . .]

Power and riches appear then to be, what they are, enormous and operose machines contrived to produce a few trifling conveniences to the body. [. . .] They are immense fabrics, which it requires the labour of a life to raise, which threaten every moment to overwhelm the person that dwells in them, and which while they stand, though they may save him from some smaller inconveniencies, can protect him from none of the severer inclemencies of the season. They keep off the summer shower, not the winter storm, but leave him always as much, and sometimes more exposed than before, to anxiety, to fear, and to sorrow; to diseases, to danger, and to death.[98]

TL;DR: We work until we're almost dead just to find out the things we worked so hard for didn't bring the happiness or fulfillment we were promised. We work to accumulate this precious and precarious status, then we spend our lives petrified of losing it. We become so obsessed with protecting it, or having more of it, we don't spend a moment enjoying this incredible life we've been given.

By the time our bodies are spent, and we realize it was all an illusion, we get to spend our old age wishing we had spent more time doing the things we love. We realize with regret that we would have been just as happy—or even happier—enjoying the "humble security and contentment" we abandoned in the pursuit of wealth. But all along we could have worked half as hard for half as long, bought half as much stuff, and been far happier. Ninety years after Keynes' essay, how many people do you know comfortably working fifteen hours a week?

Or to put it in tweet format:

We were told this story about the American dream, about the pursuit of riches, because it felt true. Because there was a time when it was true. Because leaving your impoverished village with no running water and scarce food to go to America—the land of plenty—felt like an obvious choice. The part we didn't understand was that becoming rich only worked for a minute. It worked so long as we were pulling ourselves out of poverty and giving ourselves the luxury of time, but it stopped working the moment we spent that time working even more.

If endless growth capitalism had been working to make everybody happier this whole time, you could make an argument that we needed to keep it at all costs. But it hasn't. A quarter of us are too poor, a quarter of us are too rich, and the middle half are fucking miserable. We spend close to 40 percent of our waking lives working in jobs that we mostly hate to make money for someone else and try to squeeze in a few hours of family time or alone time in between all the other shit we have to get done. We keep working sixty hours a week 'cause I'm definitely gonna be happier real fucking soon. But real fucking soon never comes.

Fetch the Bolt Cutters

(It never was America to me.)

—Langston Hughes

Hey, cheer up, buttercup! Just because late-stage capitalism is the only framework within which we've ever existed and on which our entire society depends, doesn't mean it's the only one that can ever work! Just because right now every company is only ever going to optimize their aglet production at the expense of human wellbeing doesn't mean there isn't an economic system that allows people to still make money and enjoy life without destroying the entire planet!

So we know we need a system that optimizes for human wellbeing above profit. We know the system we have is too big for the system we live within. But we also know that the economic system we created doesn't really like getting smaller. It avoids getting smaller at all costs. And above all, we know that any society willing to sacrifice human lives for the sake of protecting GDP growth has got its priorities pretty twisted.

So how do we start thinking about unwinding the overly bloated economy we've created to get back within planetary limits while increasing human wellbeing without throwing the country or the world into another Great Depression? And without, you know, sacrificing human lives. And how do we think about doing that in a way that actually makes us *happier*.

The fun part is, it's totally doable! We can make an economy that uses less energy and resources *and* people can still make money *and* people can live happier, more fulfilling lives all at the same time!

This idea has a name, and it's called "degrowth." Degrowth as a philosophy is somewhat new, but it's gaining traction. If an infinite growth economy isn't possible, and we can't sustain our current levels of resource consumption, and it's not making us happy anyway, then the only choices are to innovate our way out of resource use . . . or get smaller.

But getting smaller doesn't have to mean suffering and recession and depression. Because we've spent so long working under the growth-at-all-costs paradigm, industries have become vastly larger than they ever needed to be. There's plenty of fat to be cut while increasing quality of life for the greatest number of people.

Since when we talk about growth, we're talking about growth in GDP, there are some things we can break down that will "ungrow" it that are objectively good for the planet and for people. Maximizing economic growth in fully developed economies only works to exacerbate inequality and make people more isolated, more miserable. And all that growth pretty much just goes to making those eight white guys richer. Maximizing GDP is also essentially maximizing environmental destruction because nothing we buy comes from outside of the ecosystem. And GDP growth is pretty much one-to-one with energy usage. And we need to be using a lot less energy. We really just need to stop using GDP as a metric altogether because it's a very terrible measure of how well a society is actually doing.

But why do we keep optimizing for a thing that's making so many of us miserable in the first place? Well, because people in power really like staying in power, and getting more of it. And money too. And academic folk really hate being wrong. Soooo economists and politicians—and basically everybody—have to admit they were wrong and also give up all the power and money that these models and policies have given them. This may sound highly, highly unlikely, but the cool thing is . . . it's happening. Because some people realize that if the entire economy collapses or the ecosystem collapses, there won't be any power or money for anybody. And other people are doing it because—get this—they actually just care about other humans and our planet. It's wild!

New Zealand recently adopted a "Living Standards Framework" to replace GDP as a measure of the wellbeing of the country and her people.[99] The LSF will be used to craft economic policy in line with what's best for individual wellbeing, collective wellbeing, and the wealth of the country—with wealth including monetary wealth, human capability, and the natural environment. Pretty cool, huh?†

Another idea that's been floating around since at least 2014 is the Social Progress Index or SPI. The SPI looks not to get rid of GDP, but to complement it by measuring all the things GDP doesn't. It measures fifty-four different indicators like access to electricity, political freedom, and access to opportunity that fall generally into three buckets:

- Does a country provide for its people's most essential needs?

- Are the building blocks in place for individuals and communities to enhance and sustain wellbeing?

- Is there opportunity for all individuals to reach their full potential?

Through this lens, there are plenty of things that don't grow GDP that are still beneficial for a nation. And even crazier is there are plenty of things that would actively make GDP smaller while objectively making things better.

We talked way back about things that increase GDP while being terrible for people and the planet like smoking cigarettes, getting cancer, and oil spills, but there are plenty of other sectors ready to be unwound. Think about it for a second. What industries do you think are crazy and wasteful? What parts of the economy are making money but aren't

---

† In case you're not aware, New Zealand's wealth in the "natural environment" category is un-fucking-paralleled.

contributing to our wellbeing? That is what degrowth aims to identify. Ways to reduce overall production and consumption while *increasing* wellbeing of the general population, without the catastrophic effects of unplanned degrowth like a recession or depression.

A lot of the things I was suggesting in the last section would certainly make the economy smaller and are pretty obvious in terms of wellbeing. How much wellbeing does the average person get out of the $412 billion F-35 fighter jet program? Because unwinding half of military spending that comes close to a trillion dollars annually would have a massive impact on both the economy and our carbon footprint (and maybe geopolitics as well).

Does watching advertisements improve your life? Because banning advertising would cut $154 billion from the US economy, and that doesn't even include the savings from all the products we're not buying and energy we're not using and materials that get to stay in the ground. Creating wants we didn't have for trivial things so we can go satisfy those wants is more like brainwashing than economic development. As New Deal politician John Kenneth Galbraith said back in 1958:

> In an age of big business, it is unrealistic to think of markets of the classical kind. They set prices and use advertising to create artificial demand for their own products, distorting people's real preferences. Consumer preferences actually come to reflect those of corporations—a "dependence effect"—and the economy as a whole is geared to irrational goals.[100]

Speaking of irrational goals, does throwing away clothes you never wore or things that are "so last season" make you happier? Because fast fashion is about $106 billion, and the entire global fashion industry is $1.7 *trillion*! How much waste could we save if people didn't feel the pressure to be constantly updating their clothes to keep up with the latest trends . . . if you wore every shirt until it fell apart (or even mended it when it did) and didn't change the furniture in your living room simply because you wanted it to "look fresh." And fast fashion isn't just about materials (though the fashion industry is responsible for 92 million tons of clothes-related waste each year). Textile dyeing is also responsible for 20 percent of global wastewater (and it's toxic). It's water-intensive, energy-intensive, and I know you already know that basically none of these fast fashion brands are paying human beings a living wage. Even higher-end brands have been secretly replacing quality work with Asian slave labor for decades. Also, when you think about it, it's pretty crazy to just throw away a perfectly good winter coat (or couch) because . . . you're bored with it.‡

Same goes for cars. The automotive industry in the US generated $1.53 trillion in revenue in 2021 and accounts for three percent of the entire national GDP.[101] How much of that do you think was driven by people buying a new car because they felt like it? Because they didn't want to be seen driving their 2010 hatchback? Or people who got a promotion and thought it was finally time to upgrade from the Hyundai to the Lexus? Every year, Americans spend $698 billion on car payments and insurance.

There is a lot of GDP that is vital to our country's functioning. Like healthcare, and education, and agriculture (even though we really need to overhaul that one), and even making products that make life more convenient like blenders and refrigerators and cars. But we need less. Less consumption, fewer cars, longer lasting refrigerators.

Now let's look at the opposite side: things that are good and beneficial to society that do not increase GDP, or possibly even reduce it. How about education? Someone leaving a career in finance in order to become a teacher would reduce GDP. There is one less person making and spending $500,000 a year and one more person being tragically underpaid for very important work. If every single person left corporate finance to become a teacher (and maybe if we shrunk that industry by banning highly leveraged, risky investment practices while we're at it) there would be a huge drop in GDP. But there would likely be an increase in the quality of education, which is good. Or what about a father who quits his job to stay home and raise his baby? Bad for GDP,

---

‡ I don't want to hear about "donating" your old clothes. At least half—and probably more—of that shit ends up in a landfill. End of story.

good for baby. Or what about a city that makes a massive investment in public transport that allows tens of thousands of people to become car-free? Well, that investment would first be a boon to GDP, but then there would be a drop as all those people stopped buying gasoline and cars and paying for oil changes and maintenance. That's bad for GDP, but it's better for the environment, for the community, for the people—better for basically everyone and everything except GDP. And the fossil fuel industry.

And beyond reducing consumption or dismantling industries that shouldn't exist at all, how many people are currently working in industries doing jobs that don't even need to be done? The late, great David Graeber coined a term for an entire sector of completely meaningless work: bullshit jobs. These include the enormous gatekeeping administration for US healthcare, bloated global bureaucracies that spend all their time testing whether people should receive welfare or not, and administrators who create meta-work that obstructs workers from working—whether in the form of teachers teaching, nurses nursing, whatever.

The system is even more toxic when you realize that the most essential jobs like nurses and teachers and farmers actually pay the least. If we learned anything during the pandemic, it's that the most we're willing to give essential workers for the vital work they do is a round of applause. And then, of course, there's the other essential work that isn't paid at all. Like cooking, cleaning, raising children, and caring for our elderly parents.

In his book, which is aptly named *Bullshit Jobs*, Graeber argues that we're living in a system more like managerial feudalism. Most of the middle managers, box-tickers, and lobbyists are like the earls, knights, and other royal hangers-on. They uphold the system because they continue to profit from it, and they don't have to get stuck in one of those unlucky plebian professions like "nurse." But all of us are really just here to funnel wealth to our corporate overlord kings.

One step further from the bullshit jobs, we can find batshit jobs: jobs that actively work to destroy the planet and harm the people doing them while they do them. Like working in a coal mine that you know is destroying the mountain it's being removed from, destroying the air we're breathing, *and* destroying your own lungs while you mine it— but you don't have a choice. Everybody has to have a job because this economy better keep growing!

How much could removing bullshit and batshit jobs de-grow the economy and our environmental footprint while still not actually making anything worse for anyone except the owners of those batshit means of production who have already profited plenty from exploiting the planet and our labor? I don't know, but I'd love to find out.

Okay, I know what you're thinking . . .

But you can't just get rid of all the jobs! We need those!

The thing is, you don't have to be worried about making industries smaller when the system itself is built to support fewer people working. Everybody only works fifteen hours a week, so every full-time job that we actually need to keep could arguably be split up among three people. And don't you know how many jobs these robots are going to be doing? And the lazy people who want to sit on their ass all day and live off their universal basic income of $12,000 a year? Go for it!

But Taylor, I can't live on a third of my salary! I need my forty hours a week!

Turns out, you don't! Corporate profits have been skyrocketing for decades while we're all struggling to get by. It may not feel possible to you, but it is. Just think about all the money in America. Think about all those tech billionaires and crypto billionaires and just average hedge fund managers with fifty million bucks in the bank. Think about the private equity firms who bought every house in your neighborhood and jacked up the rent by 30 percent. There was always enough money and resources to support everyone. We just ended up with a system that funneled it all to the very top. And now we need to dismantle parts of that system to more equitably distribute all this money they made by burning ten million years' worth of liquid sunlight, replacing us all with machines, and pocketing the profits.

I know, I know, that sounds like a social welfare state. Nobody likes the term "wealth redistribution" except for dirty, dirty socialists. Everybody wants to be rich, and nobody wants their hard-earned tax dollars paying for some lazy freeloader to sit on the couch all day. But— hear me out—what if *all* of our hard-earned tax dollars paid for us *all* to sit on the couch ... but just like, half the day?

President Biden originally ran on a platform promising to raise taxes solely on earners over $400,000. More than half of America hated it. Why? Not because they made that much and would be affected by it, but because they believed one day they *might*. Only 1.8 percent of Americans earned more than that in 2019. That tax increase should have been logically supported by 98.2 percent of the country. And instead, poor people rallied against it because the idea we're all one app idea away from becoming billionaires is so deeply ingrained in our national ethos.

In order to reorient the system toward a wellbeing economy, we have to detach ourselves from our endless drive to increase wealth and productivity. We need to be okay doing less and having less. And the best part is, it makes us happier! Work sucks! They Stockholm Syndromed us into believing working too much is something to be proud of. We wear our busyness like a badge of honor. Damn, they're good.

But now it's time to fetch those bolt cutters and break the fuck out. You can still be rich if you want to. We're just getting rid of billionaires. People with hundreds of thousands of stacks of millions of dollars. If you started saving $10,000 a day around when the pyramids were built in 2550 BCE—and you never spent a single penny for more than four thousand years—you *still* wouldn't have anywhere near a billion dollars. You'd only have about $45.6 million, or about 0.02 percent of Elon Musk's current net worth. Not two percent, two hundredths of a percent. A billion is a thousand millions, and you would only have forty-five of them. Barely enough to get you into the rich people's club.

By the way, how's that going for you? Are you working on your app on the weekends? Are you one crypto trade away from hitting the jackpot? Or are you maybe too effing tired and overworked to do anything other than what you need to do to get by? How about we make it so that nobody has to worry about surviving while changing the underlying values of a system that tells us our value as humans is solely defined by how much money we make, and the people who want to bust their asses are free to keep on busting?

I know this is all easier said than done. These cultural changes don't happen overnight, but the good thing is, they already are. I absolutely love the Gen-Z/Millennial trends of "quiet quitting" and "bare minimum Mondays." We're showing our bosses and their bosses that we're not going to sacrifice our own wellbeing just so your company's stock price can go up by three pennies. We're telling them: I will do the job you pay me for and nothing more. You don't get to exploit me the way you've been doing for a half-century while you dangle the carrot of a promotion or a raise—or a sense of self-worth—that never comes.

But now we need to take it one step further. It's not about demanding more money from our bosses; it's about demanding less work for the same paycheck. It's about demanding a better life for everyone.

Above all, it's collectively recognizing that the system we have isn't working. Employers should *already* be paying you at least twice as much for the work you do—which means you could already afford to work half as much. It's a provable fact.

Back in the early days of the Incredible Everything Machine, wages rose with productivity. The more goods and services companies were able to produce, the more workers got paid. It was great. And it didn't happen by accident. It was a set of policies we had put in place to make sure that profits were shared more equally among the people responsible for creating those profits. Quick reminder: There is no economy, no country, no government without you or me. We're not "lucky" to have jobs; they're lucky to have workers. They can't exist without us.

But something happened in 1979. The powers that be started dismantling all those policies meant to ensure worker wages rose with productivity. Excess unemployment (that leads to higher competition for jobs and lower wages) was tolerated as a means to keep inflation in check. Raises in the federal minimum wage became smaller and rarer. The National Labor Relations Board (NLRB) started slacking on their pimping to rein in hostility toward unions. Labor laws were weakened, gutted, or unenforced by design. Union membership fell off a cliff. And anti-worker deregulation along with the de-toothing of anti-trust law and the dismantling of financial regulations allowed corporations to go absolutely buck wild. Here's a fun chart from the Economic Policy Institute:

The Productivity-Pay Gap
Workers produced much more, but typical workers' pay lagged far behind

Disconnnect between productivity and typical workers' compensation, 1948-2013

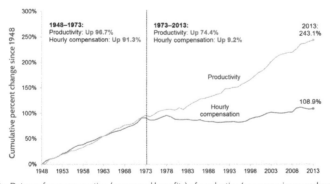

Note: Data are for compensation (wages and benefits) of production/nonsupervisory workers in the private sector and net productivity of the total economy. "Net productivity" is the growth of output of goods and services less depreciation per hour worked.

Source: EPI analysis of Bureau of Labor Statistics and Bureau of Economic Analysis data

If you work for any publicly traded company, I beg you to look at the CEO's salary and "earnings per share" and ask where all the profits you've been generating are going.

Productivity grew 8.1× as much as average worker pay from 1973 to 2013. And I think we can all agree that trend has continued—or worsened—in the decade since then. And the cherry on top? They slashed the top marginal tax rate from 91 percent in 1961 to 37 percent today. Your boss isn't just making twenty times as much, he's keeping most of it. Maynard Keynes was right. Your salary should be multiples bigger than it is for all the profits you produce. And if your salary should be three times as big, then you could afford to work a third as much as you do. I'm amazed we've even managed to survive when inflation rose 318 percent over that same period. Oh right, women were forced to enter the workforce when a single salary stopped supporting a family. And all those sixty-hour weeks we've worked have gone straight into our bosses' pockets. They literally, provably stole that from every one of us.

If you've been a human adult in America in the past ten years, I'm sure you are viscerally aware of how bad it's gotten. No one can afford anything anymore. By 2013, the ranks of the working poor had reached 47 million people in the United States—one out of every seven Americans. Back then a quarter of workers were in jobs paying below what a full-time, full-year worker needed to support a family of four above the federally defined poverty line. Today, the working poor have swelled to 58 million, and nearly 30 percent of us are living near or below the poverty line.[102] And these are *working* poor. They've got jobs. Jobs that don't pay shit. Haven't you seen *The Wonder Years*? Or *The Simpsons*? Or even *Married with Children*? Don't you remember how a family of four could be supported by a single income in a relatively unskilled job like factory worker or fucking shoe salesman?

Today, nearly 44 percent of all workers in the US earn less than $15 an hour, which I don't have to tell you, does not support a family of four anywhere in America. A full-time worker making the current minimum wage of $7.25 an hour is well below the current poverty line just for a family of two.* The average American isn't worth less than we were forty years ago. We haven't gotten lazier, and we aren't buying too much avocado toast. We lost the bargaining power needed to share in the bounty of capitalism's profits.

For the first three decades after World War II, we had that. And it was great. It was the American fucking dream. We're not all suddenly living beyond our means. The cost of living has increased beyond our means when our means should have been keeping up with the cost of living at the bare fucking minimum. As Robert Reich says, "to attribute this to the impersonal workings of the 'free market' is to ignore how the market has been reorganized since the 1980s, and by whom."

We had a system of countervailing power like unions that kept a more even distribution of money and profits and power. And even though it would be beneficial for our corporate overlords to cut us a break and accept a smaller share of a faster growing economy (because we can finally afford to live and buy more shit instead of just surviving), they're just not gonna do that. The status quo is too comfortable. Countervailing power is risky and unpredictable. Just look at how hard Starbucks and Amazon are fighting the first union push in decades. And it's only getting worse. The Supreme Court just ruled—in an 8-to-1 decision—that companies can now sue union members for financial damages caused by striking.[103] Correct me if I'm wrong here, but isn't that the entire purpose of going on strike?

Look, I'm not here to end capitalism.† I'm not suggesting we abolish private property and take away the ability or the incentive for people to make money. I'm here to suggest a version of capitalism that provides an alternative path to happiness. I'm here to argue for a future where the systems we are forced to live within are designed to maximize for human wellbeing before profit. And after you take care of all the people in your company and your community, then fuck it, go make some money. The craziest part is, this isn't even radical

---

* If you're wondering, the poverty line for a single person is $14,580 and $30,000 for a family of four. A full-time minimum wage job pays $15,080. Woof.

† Saving that for my next book.

capitalism. You could already be working two days a week if we had just kept the more equitable brand of capitalism we already had forty years ago!

As soon as we provide other avenues through which to measure value, I'm pretty sure a lot of people are gonna choose those two-day workweeks plus time with their families over sacrificing the best years of their lives for a chance to be a bit richer. I'm pretty sure that a lot of people wouldn't care as much about money if we didn't tell everyone it was the only way to be happy. As soon as inequality goes down, people care less about acquiring more. It's a fact. And when we acquire less and consume less, our economies shrink without hurting anyone. And you know what . . . not working is fun! Relaxing is awesome. But half the country can't afford to relax and the other half thinks taking a day off makes you a lazy piece of shit.

You know what most Spaniards do every day? Take a three-hour break in the middle of the day. Most small businesses close from two to five in the afternoon, and everybody eats a nice big lunch and then takes a nap. I mean, I don't, because I'm an American woman with a hopelessly ingrained Puritan-Protestant work ethic. But I'm sure it's glorious. I'm working harder on allowing myself to work less, which sounds insane now that I'm writing it. But at least I'm working hard on things that matter to me.

Okay. I'm gonna stop working and go sit on a sunny *terraza* and have a drink.

Well, that was nice! Oh, that's right. You know what else we do here? Sit on *terrazas* (restaurant patios) outside sipping on wine or beer or vermouth for hours. Slow sipping. Nobody really gets drunk.‡ You just enjoy your nice cool drink on a nice sunny terrace. No one's ever really in a rush to get anywhere. It's almost impossible to finish a lunch here

---

‡   I mean, I do.

in under three hours. You spend your time with people you love doing the bare minimum of work you have to because no one is here living to work—they're all working to live. Spaniards still work hard and still take pride in their work; they just also enjoy a lot of vacations. We walk places. We take naps. We have an incredible quality of life with far lower obesity rates and higher overall happiness.

In 2021, Spain's per capita GDP was $30,103. In the United States, it was $70,248. According to the SPI rankings we talked about earlier accounting for things like access to healthcare, water, and shelter; access to basic knowledge, information, health and wellness, and environmental quality; and personal rights, personal freedom, inclusiveness, and access to higher education, the United States came in 24th. We didn't even make it into the top tier. Spain came in 20th place. That means they are experiencing an objectively better quality of life for less than half of our GDP. Just 42 percent. Which means, it's 100 percent provably possible that the United States could have the same or better quality of life while degrowing GDP by 58 percent. That's totally nuts. That's $13 trillion of GDP that we don't even need—that isn't making us happier. As Jason Hickel says in his book *Less Is More*:

> That's $13 trillion worth of extraction and production and consumption each year, and $13 trillion worth of ecological pressure, that adds nothing, in and of itself, to the fundamentals of human welfare. It is damage without gain. [104]

Of course Spain isn't perfect, but there are boundless examples of people being a whole lot happier with a whole lot less. The list of countries with smaller economies and happier, healthier people—with more freedoms than the "freest country in the world" goes on.

In 2016, Costa Rica's per capita GDP was just $14,232 (compared with $57,866 in the US the same year) and yet is consistently rated as one of the happiest countries in the world. In that year, on the "Happy Planet Index Report" which rates countries on a blend of happiness, wellbeing, life expectancy, inequality, and ecological footprint, they ranked #1 while the US ranked 108th out of 140 countries. Yikes.

You don't have to be rich to be happy. You need to have your needs met and spend your time pursuing meaningful things. And we already have the resources in America to create a system where everyone's needs are met. Even without degrowth, just taxing investment income at the same rate as earned income would give back $1.3 trillion a decade to the people who actually earned it. How about that tax on billionaires or the estate tax or raising the social security cap? How about all that extra military spending just waiting to be unwound? We could have free public healthcare and childcare and higher education and guaranteed housing and robust community programs and public transportation and high-speed trains. We don't need a $25.46 trillion dollar GDP to do it. Europe is right here, we're already doing a lot of it, and we're already a whole lot happier. Europe needs to get smaller as well, but the point is, there is a lot of room to degrow GDP before you'll even notice it.

And when we start rethinking what work means and why we do it—when we stop existing under the same profit-at-all-costs motivator that has been so deeply ingrained in us for so very long—we can start to see what an economy looks like that is centered around human happiness.

There is a very real world where we can all be ourselves and be happy and let the inventors invent, and the lazy people be lazy, and the poets write poetry, and the scientists do science while also not destroying . . . literally everything.

It only feels impossible because we've been told it is. We've been told the second you slide into socialism, the system collapses. But there are numerous examples around the world demonstrating how wrong that is, no matter what the media tells you about Venezuela. Scandinavian countries with robust "socialist" safety nets consistently top the list of happiest countries in the world.

We've been told our Wild West brand of capitalism is the only system that works, and you'll be standing in bread lines if you don't let this CEO make $350 million a year. We've been told if we just work hard enough, we can make $350 million a year too. We've been told making $350 million a year is the only thing worth wanting. But it isn't. And in twenty years, the only people who are going to remember you stayed late at the office every night are your kids.

There are a million different ways to look at these systems and a million different levers to pull to start making things better while unwinding parts of the economic system that are adding zero value while threatening the future of our entire species.

Whether or not everything is going to collapse, we should have these things anyway. Whether or not we need to shrink the economy to save the planet, corporate profits shouldn't come at the expense of human welfare in any country, and certainly not in the richest country in the world. No one in the richest country in the world should be struggling to get by. No one in the richest country in the world should be starving to death or freezing to death on the streets. And no one who's already making $80,000 a year should feel like they're poor.

And when we all stop maximizing for a thing that makes us miserable, maybe we'll all get a little less miserable?

By Design or Disaster

**Will mankind listen to any program that implies a
constriction of its addiction to exosomatic comfort?
Perhaps the destiny of man is to have a short but
fiery, exciting, and extravagant life rather than a
long, uneventful, and vegetative existence.**

—Nicholas Georgescu-Roegen
Energy and Economic Myths

We can reduce the energy throughput of our economies without
decreasing happiness or wellbeing at all—or even increasing it in many
cases. We should be doing that whether or not we're facing an imminent
crisis. Of course, we are facing imminent cris*es*. Whether or not you
find the ideas behind degrowth inspiring, and whether or not you're
thinking about quitting your job and moving to New Zealand, it doesn't
really matter. Slice it any way you want it; there are pretty much only
two options ahead of us, and both of them involve degrowth. One is
degrowth by design, and the other is degrowth by disaster.

We can choose to make changes that will soften the blow when the
unsustainable systems we've created start to unravel, or we can just sit
here and watch as the world burns. Depending on who you are and how

optimistic of a person you are, you'll fall into one of these camps. But the hard truth is: there is going to be some kind of collapse. I just don't know which kind or when. Could it be oil prices or geopolitical shifts collapsing globalization? Could it be the collapse of overleveraged economies that can no longer service the insane amounts of debt they've taken on, causing a ripple effect in global financial systems? Could it be a French-revolution-style uprising when the inequality and oppression in America *finally* go too far? Could it be the second civil war in America? Month-long heat waves that kill hundreds of thousands of people? The Amazonian hydrological pump breaking? Full-on ecosystem collapse? All of the above?[§]

No matter which future we face, whether hopeful or dystopian, there is one reality they all share: energy descent. There is no future where the world can continue to use the same amount of energy and materials we've been burning through for the past seventy years. And when we start thinking about a world with far more limited and far more expensive resources, we should probably figure out a plan for where they should go.

For now, you're happy to get in your car and run to the store whenever you need something, but imagine that with $25-a-gallon gas and $2,000 electric bills every month. How many appliances would you start to unplug? Would you keep pressing the button that opens the door on your minivan if it cost five bucks in gas every time you pressed it? How many people would move to areas that were more walkable or bikeable when the cost of driving your car around increases by 500 percent?

Politicians suggesting massive public transportation investments aren't very popular in a country where people are addicted to their cars, but that might look a whole lot different when people are adjusting to a reality where they can't afford to drive.

Universal basic income might not be very popular in a country founded on rugged individualism (and kill-or-be-killed capitalism), but that might look a whole lot different in a country with 40 percent

§   Don't forget to rip out your Bingo card, folks!

unemployment because the robots took our jobs, the CEOs took the profits, and we're facing widespread food scarcity and homelessness in a landscape that's starting to look suspiciously like the Hunger Games.

If we do nothing at all, the markets will take care of things to some extent. But it won't be pretty. Goods will become prohibitively expensive for most people, and we'll be forced to return to simpler lifestyles. But if we just let the market handle it, we're gonna have rich people continuing to live the lives they lead—at a much bigger price tag—and poor people suffering the consequences. To be honest, it's already happening. And I don't mean Africa poor or South Asia poor. I mean America poor. If you don't have the money to spend $6,000 on a cell phone or pay $50 in gas just to drive to the grocery store and back, you don't get to.

If we don't figure out a way to better manage these resources before it's too late, they're just gonna go straight to the one percent like everything else has for the past forty years until one day all the billionaires are living in a floating utopia above the Earth while the rest of us plebes starve to death on a scorched, barely inhabitable planet.

I want you to really imagine that future. Okay, maybe not that one. But think about a future where gasoline is too expensive to put in a car. If it's too expensive to put in a car, is it too expensive to make an electric car as well? I guess everybody's riding an electric bike. Hopefully cultural materialism has precipitated a change in our transportation infrastructure to make that version of life more livable. But there's a better-than-good chance we don't get there in time.

This is why we can't just let it shake out and see where things land. This is why we can't rely on the innovation of capitalism to solve all of our problems when no one right now believes these problems are urgent enough to need solving, so there's no market incentive to do anything at all. This is why we need to have local governments working to create a world that doesn't suck for everyone who's not already a millionaire. Maybe we should start planning ahead just a bit and start to create a society that's a little more resilient than our "the second you're fired you lose your healthcare and good luck getting on food stamps you lazy piece of shit" model we're working with now.

Maybe rather than just demanding a higher minimum wage or more vacation time, we start rethinking the purpose of government in our lives and society. Maybe it's not just to provide basic services like roads and trash and driver's licenses (and healthcare), but to manage this public wealth and use it to provide all the services that are essential to living a fulfilling life: public transportation, for sure, but also things like affordable housing for everyone, community programs, playgrounds, open areas, parks, libraries, investments in arts and culture. Sure, we already do these things to some degree, but we need to do them better and more. We need to do them *completely*. We need to look to government and corporations to take all that money they've been hoarding and use it to help rebuild and redesign communities around what's best for the people within them while preparing for a future in energy descent.

This is what a wellbeing economy is all about. It's the Living Standards Framework. It's making active, conscious decisions directed at increasing wellbeing for the most people. It's making policy choices not on what will grow GDP the fastest, but on what will shrink GDP, because that's what's better for the planet and better for humanity. And in addition to governments working to actually improve our lives, we need to start thinking about how governments should craft policy for corporations to increase social good as well, just like Adam Smith envisioned.

We already talked about stakeholder capitalism and B Corporations (and paying humans a living fucking wage) but what about corporations 2.5? What if every business was required to take on a particular environmental or wellbeing initiative like carbon capture or improving transportation or public arts centers or early childhood education or cleaning up the Great Pacific Garbage Patch? What if every business was forced to reinvest a healthy portion of their profits directly into the communities they couldn't exist without or the problems their business models contributed to? What if we taxed billionaires and corporate windfall profits so obscenely that our government had money to give us literally everything we need to live happy, healthy, wealthy lives? The real wild part is we are completely capable of doing all these things while degrowing the parts of the economy we don't need and focusing on the parts that we need more of. We have to allow ourselves to imagine a better future because it's 100 percent possible.

Water, nature, land, and food should be public goods. The lithium, the cobalt, the oil . . . How is it that a company is getting to extract those things underneath our feet, keep the money, price gouge us when they want to, and we get nothing? Do you still want Exxon owning all the oil when it costs $800 a barrel? Do you want Nestle owning all the water when it starts costing $25 a gallon?

The minerals in the Earth belong to everyone. And if we were all—at the very least—sharing in the benefits of exploiting those while they last, most of our slices of pie would get a whole lot bigger without growing our economies or using any more energy at all. But more importantly, the pie would get a whole lot more equitable. Imagine the wealth of your country, everything under the ground, as an inherited resource. Imagine if the $200 billion in profits that oil companies raked in in 2022 belonged to the people who lived on top of that oil—the people who were impacted by the earthquakes and poisoned water caused by the fracking. Every single year, every resident of Alaska gets a $1,600 paycheck for the oil that gets taken from their stunning landscape. How come the residents of Texas don't? Or Louisiana?

It won't save us from depleting it, but fuck me, at least we'll have the dignity of security during the greatest period of energy abundance our species has ever seen.

Now I want you to think about the next step. I'm not just talking about gas prices going way up. I'm talking about maybe there isn't enough gas to keep the machine running. This isn't a potential future; it's a definite one. It's just a question of when. If oil is going to run out in just fifty years—and we don't have anything to replace it—what does that future look like? Or even if it lasts another hundred years, what does the future look like then?

When we start running out of oil, do you think we're still gonna use it to gas up a private jet? Or build an NFL stadium? Or will we save it to build hospitals? Will we use it to mine for more rare earth metals so we can make more of the energy infrastructure we need to replace the oil? And what about all the metals we can't make and refine without oil? If we use steel now to make Lamborghinis and corporate skyscrapers, there will be less in the future for more important things like making food and building houses.

There's a concept in economics known as diminishing marginal utility. Diminishing marginal utility tells us that when you have money, you're going to spend it on the most important thing first, and then the next most important thing, and so on. So first you'll buy food and pay your rent, and lastly, you'll buy Gucci sunglasses (hopefully). If we know our resources are finite, and our goal is "the most happiness for the most people" then how do we allocate the resources we have left to derive the maximum utility?

Part of the answer to degrowth by design, to soften the blow of resource scarcity, economic instability, and rising prices, is top-down. We need the underlying infrastructures to change. We need the rules of the game to change. We need the government to take action to help more people. We *already* need the government to take action to help more people, and none of this has even happened yet. We need all those high-speed trains and purpose-built communities, and legislation that focuses on human wellbeing instead of endless profits. But there is another piece of the puzzle that could help soften the blow . . . and that's from the bottom up.

Remember earlier when we were talking about exosomatic energy? As a reminder, in America we use about 300 gigajoules per person per year, or 100× exosomatic energy compared to the energy we use to feed our bodies. But what's interesting is that number isn't remotely the same across different industrialized countries. American life is an anomaly in the anomaly. Once again, let's use Europe as an example. I think we can all agree that Europe enjoys a quality of life at or above the standard in America. We've certainly all got clean water and indoor plumbing and culture and art and parks and good education and healthcare and movie theaters and lots of grocery stores with lots of products (though not forty thousand). People over here are pretty happy. Europeans work less on average and take more vacations, and we have Eurovision! We've already shown that you can have a great quality of life over here with a much smaller GDP, but how much energy are we using to keep that up?

Welp, the thirty-three countries I pulled from in Europe* had a total population of 659.5 million in 2021, and according to BP's annual global energy use report, used a combined 80.66 million gigajoules.[105] That's about 122 gigajoules per person. Less than HALF of what America takes to live a fully developed, comfortable, industrialized lifestyle. Hmm.

What do you think makes up that difference? Well, Europe on the whole has wayyyyy better public transportation and urban design. People don't own cars because you don't need to, and suburbs are designed to still be walkable or accessible by public transport, as opposed to suburban America that was basically urban planned by the fossil fuel industry.

Public transportation is an absolute abomination in America. First of all, taking the bus in most cities carries a cultural stigma that you're poor even though that's absurd. But they're also not wrong because the buses in most cities are so unimaginably terrible you would never take them if you had a choice. Most city bus systems can turn a twenty-minute drive into a two-hour ordeal. And when you rely on the bus, but the bus is unreliable, then you're late to work. And of course, your boss doesn't give a shit why you're late, so then you get fired. And then you miss rent so you get evicted, and then you can never rent another apartment because every apartment complex is owned by massive corporate leasing companies that don't give a shit about you. And good luck getting any welfare benefits if you don't have a job in America because that makes a

---

* In order to keep my data apples-to-apples, I'm looking at 33 European countries that contain most of the continent's population. BP data lumped together everything from Albania to Moldova into "Other Europe," and I wanted to be sure I was directly comparing population to energy use.

whole lotta sense. Please read the book *Tightrope* if you want a visceral and moving portrait of cyclical poverty in America. </rant>

No one in America would take the bus if they didn't have to. And walking in America isn't only difficult—it's dangerous. You get in your car to go anywhere, and that's that. Don't even get me started on how frightening a long-distance Greyhound can be. But if we had a better system, more people would use it. And if we have great public transportation and bike-friendly cities, we don't need to make two billion electric cars with all those little children in the Congo mining for cobalt.

But there are other things too. Almost every American home has a washer and a dryer. But, I don't know if you're aware of this, clothes actually dry . . . by themselves. It's a wild combination of wind and solar power called putting them on a drying rack in the sun. And here in Madrid where it's always sunny and the humidity is never above 50 percent, in the spring and summer they dry as fast as a dryer works. All for the bargain cost of free. Most American homes have central heating that people keep between 68 and 72 degrees throughout the year. Most European homes do not. Sometimes you're hot, and sometimes you're cold. It's wild!

And on top of our thermostats and our car-centric suburbs, there is so much energy that gets used to support our lives that we don't even think about. It's so far removed from the thermostat dial as to be completely invisible. We're all guilty of this one. Think of every product you've bought in the last year. Think about its supply chain. Where were the raw materials made or grown or mined? Where was it assembled? How many times did it cross an ocean to get to you? If it was made entirely in the US (which is highly unlikely), where is the factory it was driven from? Think about those numbers as they relate to all the food in your fridge and the different continents it came from. Think about the fertilizer that was needed for the crops. Think about alllll those fossil fuels and all that needlessly spent energy.

As we start to look towards a future in energy descent, we have to think about all the energy we use that isn't pulling its weight in happiness. We have to unwind industries that don't make sense, but we also have to unwind our behaviors that don't make sense. Sure, we'll still use washing machines for our clothes because it saves us an entire day of labor. But the dryer? Nope. All it does is make your socks and towels softer.[†] Driving everywhere is terrible for happiness, for the planet, for communities. It's expensive and alienating and frustrating and polluting. Admittedly, overhauling that in a country addicted to driving is gonna be tough. Until we don't have any other choice.

I know this all sounds preachy and sanctimonious, and I wish there was a way around that. But there isn't. We all need to get better, myself included. We all need to start looking around the world and at our own lives and take note of the energy that we've only been using because we never had to think about it before. There is plenty of fat to trim while still living an incredibly comfortable and luxurious life.

We can wait until energy descent becomes a reality (which is 100 percent what is going to happen at some point without some monumental technological innovation) or we can start to shift our usage now towards the things that provide the most value. And we can stop wasting it on things that are arguably making the world worse.

As we start to talk about reducing our energy footprint in America and Europe, there is an elephant in the room I haven't yet addressed: and that's all the people still living in actual, abject poverty in the rest of the world who deserve clean water and electricity and heat and indoor plumbing that will involve further increasing the energy throughput of their country's economies—the countries where increasing GDP still actually works to increase happiness.

How much energy do they get to use?

When you apply diminishing marginal utility to global resource allocation, we can see that extra money (or energy use) going to a poor country (provided it doesn't go to the corrupt elite) represents

---

† I realize some very humid or rainy places like New Orleans or Seattle or Amsterdam might have some things to say about this. But I promise, your clothes will still dry.

satisfaction of relatively basic wants like food and shelter, whereas extra money flowing into the US (provided it doesn't go to the poorest people who still need food and shelter) will satisfy trivial wants. The people in Kenya will buy bread. We'll buy nicer cars and Gucci sunglasses.

We can't just tell all of Africa, oops, you missed the energy abundance train. Too bad, we're fresh out. And if we don't help them develop in the right way, they're just going to burn a whole bunch of coal to do it the wrong way. Just like we did when we first got started. Those 1.2 billion people in the world who still don't have electricity should be allowed to have electricity, don't you think?

I'm sure your first thought is "Well, tell those billionaires to stop flying private jets all over the place!" And you're absolutely right. Private jets should be banned. Flights under two hours should be banned[‡] (I can't stress enough how much we need better trains). But the uber wealthy aren't responsible for the lion's share of our emissions and energy usage. I mean, their per capita usage is an abomination. But it's the 630 million people (around 8 percent of global population) living extremely comfortable lives in highly industrialized countries who accounted for 52 percent of emissions from 1990 to 2016.[106] And if you live in America—even if you're poor by American standards—I assure you, you are filthy fucking rich from a global perspective. Do you make more than $38,000 a year? Congratulations, you're in the richest 10 percent of the global population. Add in the 2.5 billion humans who live in upper-middle-income countries with rapidly growing energy usage (think China, Thailand, Argentina) and that accounts for 93 percent of global emissions. Yikes.[§]

There aren't really any arguments to make that the resources are ours to keep, or the energy is ours to burn, because basically all the wealth Western countries have accumulated was possible because we took the most fossil fuels out of the ground. We burned the oil, we caused the pollution, our corporations became multi-national conglomerates with trillion-dollar market caps on the backs of workers in developing

---

‡   Hooray! Back to banning shit!

§   I know, I wish we could have blamed it all on the billionaires, too.

economies, and all because we were the first ones who had oxen to domesticate.

So what does a world with more equitable energy distribution look like? How much less do we have to have so that everyone else gets a chance at a normal life?

There is no way to frame these arguments that doesn't involve people in highly industrialized countries living with less. But the most critical piece to understand is that living with less stuff doesn't have to mean living with less joy. You can call it "voluntary simplicity" or "revolutionary austerity." But the better way to think about it as you maybe start to make changes in your own life is "living simply so that others can simply live." If you care about the billion people on Earth still living in poverty who don't have clean water or the other billion who are subjected to working in sweat shops, then we have to have less so they have a chance at having the bare fucking minimum.

The messed up thing is, I already know that argument won't work. I know nobody really cares about the billion people already starving. They've been starving for decades. They've been starving forever. Most people already know all these things and haven't changed a damn thing. You're going to keep your house at the temperature you prefer and dry your clothes in the dryer because it's easier. And asking really nicely for everyone to give up airplanes and hamburgers is a surefire way to get them to not do that at all. If you're already grabbing your car keys to go stockpile toilet paper and ground beef, just chill out for a second. Fight that feeling for just one moment and think about it. Think about all the things you have in your life. Because the most incredible thing is that most of us agree that our massive economies are big enough to provide enough to everyone.

No one doubts that America is rich enough to easily feed and house and educate all 337 million people who live there (and still have plenty to give up to allow developing economies to grow). The only argument being made is that we shouldn't—that it's the individual's job to make their own way in the world. And that it's our God-given right to become billionaires, even though none of us ever will be. And, I gotta be honest, I don't even think being a billionaire is all it's cracked up to be.

But if you ask most people what they need to be happy, their list is pretty short and pretty obvious: good food, a warm house, a healthy family, reliable transportation, a good school for my kids. People want the economic freedom to not worry about providing necessities, and then a little bit extra to treat themselves from time to time. To take a vacation or have a nice dinner out. Seriously, without the manufactured demand of our corporate overlords, that's about it.

How much do you think that baseline of happiness equates to in America? And if everyone in America was living at that comfortable standard, how much excess GDP would there be? How much excess energy would there be to go around? I'm not suggesting we all give up everything and move to the forest. I'm not suggesting we live by candle-light and cancel the internet and go back to some magical happy time before the first oil well was drilled. That time doesn't exist. Shit sucked back then.

I'm suggesting that if we don't start making changes, they're gonna get made for us. And unless you're already disgustingly wealthy, the scales aren't likely to tip in your favor. By design or disaster means it's happening whether or not you want it, whether or not you believe it. But if we start looking at our lives, there are so many answers, so many easy solutions to help our system (and ourselves) bend toward a new reality before we break under the pressure.

Enough is Enough

*Shall we always study to obtain more of these things,
and not sometimes to be content with less?*

—Henry David Thoreau
Walden

You probably didn't like reading the end of the last chapter. Or even reading the title of this one. Oh wait, I changed the title. But it used to be called "Getting Satisfied with Enough." Didn't like it, did you? Most of us Americans are so obstinate and independent that we'll keep doing something we don't even want to do for the sole reason that you told me I couldn't do it. There's talk about banning gas stoves to stop pollution right now and people are literally just turning on the burners in their house—filling their homes with toxic gas . . . to prove a point?

But I think we can all agree that's not really the way to be. We need to take stock of the lives that we're living. If we want to be the happiest versions of ourselves, we need to clean out the inventory. This isn't just about more parks and community spaces and better public transportation. Those things are great, and the changes in society need to come from the top down and the bottom up, sure. But they also need to come from the inside, out.

We need to understand all the things that occupy our lives that don't bring us joy. We need to Marie Kondo the whole damn thing.

We have to get back in touch with the things that make humans happiest. We have to get satisfied with enough so we can spend our time seeking out the things that matter most and give us meaning and joy. Otherwise, even if we're only working fifteen hours a week, we'll still be desperately unsatisfied, looking to fill that time with ever-more consumption.

Did you read that Thoreau quote from *Walden* at the beginning of the chapter? He wrote that in 1854. He was criticizing society for being too material-driven and missing out on the joys of life—five years *before* the first oil well was drilled. So obviously, we can't blame all our problems on oil. If we had already lost touch with it in 1854, we sure as shit don't have a handle on it in 2024.

There's absolutely nothing wrong with buying things and enjoying them. But the buying of the things has become the end in itself. Wealth isn't the problem; it's our attachment to it. Money isn't the root of all evil. The *love* of money is the root of all evil.

Back in 1884, William Morris, a British textile designer (who was, admittedly, a big fan of Karl Marx) said:

But think, I beseech you, of the product of England, the workshop of the world, and will you not be bewildered, as I am, at the thought of the mass of things which no sane man could desire, but which our useless toil makes—and sells? [107]

—William Morris
"Useful Work Versus Useless Toil"

As Herman Daly says, the object of economic activity is to have enough bread. Not infinite bread, not a world turned into bread, not even vast storehouses full of bread. We have all these products that 'no sane man could desire,' but we buy them anyway. We buy clothes for our dogs, and bacon-flavored soap, and light-saber chopsticks, and putting greens we can use while we poop, and blankets that look like tortillas, and harmonicas that look like pickles, and SNUGGIES. We buy all these things because it's normal and fun and we've got all this extra money lying around (I guess?). Even though most of us absolutely know in our hearts that buying shit will not make us happy, we're suckers for it. We're easily manipulated evolutionary idiots chasing whatever they tell us we should chase.

It's like we're really, really thirsty, so we keep eating Saltines. Because someone told us that Saltines would quench it. We keep shoving more and more crumbling crackers into our parched mouths—even though we're pretty well aware this isn't going to work and all we need to do is put down the crackers and take a sip of water. The Saltines are our endless accumulation of material goods. And the sip of water is all those friendships and relationships and nature and love and free time and all the things that actually make us happy.

The reason we buy all these things and keep eating Saltines isn't because we love things (or Saltines), it's to find that elusive joy. We don't inherently value the things. It's because we're told the things and the house and the cars and the yachts will do the trick. It's the craving for social status and belonging we also think will bring us joy. It's the same pieces of human nature that Smith was so adept at discerning, except trapped in a system made to amplify the worst parts of everything.

Perhaps you're familiar with American economist and socialist Thorstein Veblen and his concept of conspicuous consumption. In his 1899 book, *The Theory of the Leisure Class*, he argued that humans accumulate goods just to show them off to other people. We show off the expensive things we buy because we believe it will give us access to the next highest level in society. But this is still based on the value we have ascribed to those things. There is no inherent value in Gucci sunglasses. And in a society that doesn't value Gucci sunglasses, they're a worthless piece of plastic. Advertising, of course, plays an inextricable part in this. Once we identify a brand as a luxury brand (because they told us so and started charging $400 for a piece of plastic that cost a buck fifty to produce), we then know that purchasing it will tell other people that we are part of this social class with which we want to be associated. This entire phenomenon is just a result of over-accumulation from industrial capitalism. The whole *nouveau riche* of the early 20th century (the first people to benefit from 'rags to riches' capitalism) had to show everyone how rich they had become. Obviously, John Stuart Mill had something to say about this too:

> I know not why it should be matter of congratulation, that persons who are already richer than any one needs to be, should have doubled their means of consuming things which give little or no pleasure except as representative of wealth.

We spend money getting things just to show other people that we have money to get things. But if we hadn't created a system that allowed for the massive accumulation of wealth to a tiny social class at the expense of literally every other human in society, we wouldn't have created the desire for people outside of that social class to go into debt to buy a Porsche or a handbag. The worse inequality is, the worse this gets.

The saddest part is, the people more likely to engage in conspicuous consumption aren't the wealthy—it's people in lower social classes with something to prove. It's people who have dug themselves out of poverty just enough who are more likely to waste their money on designer clothes and car payments they can barely afford that ironically impede their ability to accumulate the kind of wealth that could actually lead to financial freedom.

But maybe you don't do that. Maybe you say, *I don't buy designer clothes, I just live a normal life.* But look around. What's normal? Even if it's not designer clothes, how much stuff do you have? And how happy is all your stuff making you? The thousands of dollars of clothes in your closet, the car in your driveway, your brand-new iPhone? I mean really look around your house and your life. You don't have to spend thousands of dollars decking yourself out in head-to-toe Louis Vuitton to be a victim of consumption culture. We are all—without exception—victims of consumption culture.

How often do you wish you had more money? How often do you think things like, *If only I could afford a nicer house or a bigger apartment.* How often do you wish you had something your neighbor or friend has? How often have you felt embarrassed because your shoes weren't nice enough or you couldn't afford to fix the dent in your car? How much do you buy because you really want it and value it and how much is to present a certain image to your friends, neighbors, and coworkers? And when you did buy the new shoes and get the bigger apartment . . . did it work? Did it make you happy?

According to a 2019 survey from the Bureau of Labor Statistics,[108] the average American household income was $82,852, which supported an average family size of 2.5 humans. Of that average household, $1,723 per month went to housing including $338 for utilities like electricity and water, $680 went to food, $895 went to transportation (including $366 for a car payment and $174 in gasoline), and $432 went to health-care. And then another $432 every month went to miscellaneous things like entertainment, personal care, tobacco, buying books, whatever. And then people spent an additional $157 on apparel.

This may look pretty normal to you. But nobody needs $157 worth of new clothes every month. Nobody. Think about that $366 car payment. That's an average that includes people with no car payment. Plenty of people buy a $70,000 car they can't afford with a $1,000 monthly payment so that their neighbors think they're richer than they are. And over $400 on miscellaneous stuff? What does that even mean? Consumer debt in America rose to $16.38 *trillion* in 2022 up from $15.31 trillion in 2021. We're not buying things we can afford because we have extra money and buying stuff is fun. We're buying things we can't afford because we think we'll be miserable if we don't.* Credit card debt grew 16 percent in a single year—more than a trillion dollars of things we couldn't afford. We're all keeping up with the Joneses while the Joneses are keeping up with somebody else.

What about you? How much is your car payment? How much do you spend on gas every month? How much do you spend on "miscellaneous" every month? How big is your house? How much do you spend on heating bills in the winter and A/C in the summer? How often do you buy new clothes? What do you do with your old clothes? How many things in your house run on fossil fuel energy instead of old-fashioned human energy? Do you dry your clothes on a drying rack or in a dryer? Do you let your hair air-dry or do you use a blow dryer?

When you start to think about how much stuff you have and how much it costs, those time prices can come in handy. If that Louis Vuitton bag cost $1,500, how many hours did you have to work to buy it?

Let's say you make half that average of $82,852 a year (because that's a household number), which comes out to about $20 an hour if you work forty hours a week and take a paltry two weeks of vacation. That's seventy-five hours of work—nearly two entire workweeks—just to have a fucking bag? So that other people know you could afford to spend $1,500 on a bag? Guys—it's a bag. It doesn't do anything different than any other bags. It's not a car with leather seats and a more powerful engine. It's just a bag.

What about your car? The average monthly payment for a new car in America is $716. That's nearly an entire week of work—a quarter of your paycheck. And if you're leasing it, you don't even get anything at the end of it! What about your house?† How many *years* of your life did your house cost you? How many more years left of your life until you actually own it? The average mortgage takes thirty years to pay off. We all think that sounds normal, but is it? Is it really normal to spend the bulk of the best years of your life just to have a very nice roof over

---

* Or we're buying necessities we can't afford because we're not getting paid enough to afford to live in a world of falsely inflated prices built to deliver value to shareholders.

† Lol who can afford to buy a house?

your head? Would a smaller roof suffice if it gave you back ten years of your life? And why do we all buy the biggest house we can afford with the mortgage amount we've been approved for? Because we love displaying our wealth. Because it's absolutely inescapable. If you live in a small house in a bad neighborhood, people will assume you had no other choice. Trust me. I've done it plenty of times. People think you are absolutely insane if you choose to live in a run-down shack in a shitty 'hood because you value saving money more than spending it.

And what about the thousands of other things we buy without ever thinking about it? All those things that cost ten bucks to get shipped from China? The art and the throw pillows and the stupid kitchen gadgets and the things we use once or twice then throw away? I'm not saying you shouldn't have art or throw pillows or kitchen gadgets. I'm saying we need to ask ourselves how much we really need and how much happiness or utility those things are providing. Because every item you buy is something you're working for. Something you're spending one hour or ten hours or sixty hours of your life to make the money to pay for.

The law of diminishing marginal utility is useful for talking about how we spend our money, but it's also useful in thinking about how much value or satisfaction we get out of things. As a person consumes an item or a product, the satisfaction (or utility) that they derive from the product wanes. The more you consume it, the less you enjoy it. For example, you might buy a certain type of chocolate for a while. But eventually you kind of get tired of it, and so you start having sour gummies instead. Or a cookie, whatever. The reason you grew tired of the chocolate is its diminishing marginal utility.

This diminishing marginal utility is true of everything in our lives. It's true of Instagram and TikTok and your car and that sweater you're wearing, and just everything. Once the novelty wears off, you look for something different to fill that gap.

> Americans find prosperity almost everywhere, but not
> happiness. Desire for well-being has become a restless,
> burning passion which increases with satisfaction.
>
> —Peter Whybrow

The more our desires are satisfied, the more we desire. And the more we desire, the less happy we are. It's super messed up, huh? This also plays into a concept known as the hedonic treadmill. I'm going to include a passage from my first book here because I'm lazy and I already wrote it:

> There was a study done a few years back (which I can't seem to find) that asked people to rate their own happiness over time, on a scale of 1 to 10. Most people, on average, would give themselves a 7. Not too shitty, not too ecstatic. Then, maybe you get a promotion and a raise, and you rate yourself a 9 for a little while. But the interesting part of the study was that over time, everyone ended up back at 7. The wonderful new apartment you got just becomes your regular apartment, and then you find something else to complain about.‡ The guy who went down to a 3 after he got fired went back up to a 7 too.

This is known as the "hedonic treadmill," a term coined in a popular 1978 study that tracked the happiness of lottery winners and accident victims who became paraplegics.[109] While the amputees were ever-so-slightly less happy than the control group, the lottery winners weren't any happier *at all* than the control group. Certainly the feeling of winning the lottery was an ecstatic, life-altering high ... until ... it wasn't.

As soon as you achieve anything on your list, it becomes your status quo. As soon as you move into that mansion, you forget all about your studio apartment. This is the new status quo, and

‡   If you've ever watched the show *Succession* you can get a nice glimpse at how meaningless fancy things are when you're a billionaire.

it fades into the background just like everything else. And then we start looking for new things that get us back up from a 7 to a 9. And herein lies the root of the problem: so long as you are looking for something else or waiting to attain something else in order to be happy, you will never be happy. [110]

The only way to actually be happy . . . is to be happy with what you already have, no matter what that is (provided your basic needs are met, of course). This approach to happiness has been practiced by plenty of "less-advanced" cultures throughout history. It's based on having few material wants that are easily met with limited energy, technology, and not too much effort.

In his book, *Affluence without Abundance,* James Suzman explores how hunter-gatherer societies were actually happier than we are with all the amazing conveniences of the Incredible Everything Machine. How? By the simple expedient of not desiring more than they already had. According to Suzman:

> Hunter-gatherers were content because they did not hold themselves hostage to unattainable aspirations. With a knack for coming up with catchy phrases, Sahlins dubbed hunter-gatherers "the original affluent society" and referred to their economic approach as "primitive affluence." The idea of being satisfied with what was ready to hand contrasted starkly with the American dream of the 1950s—a dream that celebrated the ability of capital, industry, and ultimately plenty of good, honest hard work to narrow the gap between an individual's material aspirations and their limited means. In the idiom of the counterculture movements that were sweeping through the United States in the 1960s, Sahlins characterized hunter-gatherers as the gurus of a "Zen road to affluence" through which they were able to enjoy "unparalleled material plenty—with a low standard of living." [111]

In our hyper-rich country, we look to wealthy, powerful people thinking how poor we are in comparison. We wonder why we don't have a Porsche or a penthouse or a Bentley. But what we really should be doing is asking ourselves, how did I get so rich? When just eighty years ago, more than half of America was still shitting in a wooden box outside? Being poor in America doesn't mean living on a dollar a day with no running water. It means a lot of other terrible things, but it's a whole lot closer to the baseline for human happiness than plenty of other countries around the world. And as much as we tend to forget it, we really, really need to remember that right now.

We have to ask ourselves, what is that baseline for happiness in our own lives? What is the level of material wealth above which we don't experience additional wellbeing? When do we start seeing diminishing marginal utility of money and energy use? How much shit would we actually buy if advertisements weren't telling us to buy it all the time? If we weren't constantly proving to our friends and neighbors that we've finally made it? How much excess are fully industrialized Western countries really reveling in?

How much do we need just to be happy?

The market we have is built to keep us miserable, keep us stupid, and keep us working. But it can change. Societies change all the time. Cultures shift and adapt to what's best for the survival of the people within them. The new market has to be a cultural shift that changes the values that incentivize our behavior. For most of us, the most authentic expressions of ourselves still live within the constructs of the materially focused society we've created. We can't see a version of ourselves thriving outside of the system that's defined who we are for so long. But beauty, wealth, thinness, material accumulation—none of those are inherently valuable.

The people in Costa Rica are happier on $14,000 a year because they've created a culture that values the right things. When everyone stops caring what kind of car you drive, everyone will stop buying new cars to impress people.

Much like the original market incentivized everyone's behavior to make more money to be better off (which would inadvertently benefit social good), we need a new set of underlying values that incentivizes behavior toward buying less and living more.

But here's the fun part. You don't have to be driven by a sense of altruism or a fear of ecological disaster. Because even if that wasn't on the table as a very real possibility, we still wouldn't be happy. Even if you don't really care about those people starving all over the world or the collapse of capitalism as we know it or the destruction of the ecosystem (which is a little messed up, by the way), getting satisfied with enough will make YOU happier.

Fuck revolutionary austerity, let's call it radical abundance. Let's call it alternative hedonism.§ This isn't a threat of individual renunciation: give up your comfy life or you're a terrible person. It's a promise that getting rid of all the shit you don't need will actually make you happier. It's actually the BEST way to be happy. It works.

If a consumerist lifestyle isn't bringing us the endless happiness it promised, let's find new forms of desire. Let's remind people of the shit that actually makes us happy. Let's ban the advertising that tells people buying $700 shoes will make them happy and attractive and desired and replace it with PSAs telling you to call your best friend or go take a walk in the park. Our culture in itself is a form of domination. They're hijacking our happiness to line their pockets and convincing us it's all our fault we haven't gotten there yet.

And the answer is absolutely chock to the brim with trite clichés. Money doesn't buy happiness; you can only find real happiness from within, whatever. But they're cliches because they're true. Because every single person who gets asked on their deathbed what they regret and what they loved the most in life never utters a single word about making more money.

Real wealth is wealth in relationships. Real happiness is the joy of being in love or playing with your dogs or a day at the beach or on a hike. This isn't just some mumbo jumbo environmentalists say to convince you to give up your SUV . . . it's literally, factually, provably true.

Think about all your best memories. What do they have in common? They are probably times you spent with friends or family, times you saw some incredible nature—like the Grand Canyon or going on safari if you've been blessed enough to have that experience. It's times you felt

---

§   I did not invent this term, philosopher Kate Soper did.

incredibly connected with the world around you. Whether that world was nature or an animal or another human being. Because ultimately that is our nature: we are part of the world, the planet, and everything around us. And when we feel connected to it and to others are the times we're the happiest. The things that make these lists are never "when I bought my smart fridge."

When you realize that the house and the car and the clothes will never make you happy—when you realize that once your basic needs are met, happiness is something that comes from human connection and connection with nature and being a part of something greater—then you can start living your life in a way that maximizes for those things.

This is the shift that happens from the bottom up. Much like our top-down ideas where policy impacts corporate behavior which impacts what we do and how we consume, this is the other half of the equation. When we each start making these choices, it ripples upwards through society.

As Paul Ehrlich says, we have to find ways to value human beings and their activities that aren't measured in terms of monetary exchange. You're not better because you make $200,000 a year and I only make $40,000. Looking down at the garbage man or the janitor or a bartender or a waitress just makes you a dick. The thing that we should be valuing in life, rather than financial capital, is social capital. If we want to make some hierarchy of who's winning at life, it should be the person with the richest, deepest social bonds. The person who's happiest, healthiest, and gives the most back to their community.

And just like living with less makes us happier, so does giving more. It's so funny because if you think about it for one second, I know you already know that it's true. Every single one of us already knows it to be true. And we've always known it.

So what's stopping us from doing it?

Innovating Our Way Out

Our tools are better than we are, and grow better faster than we do. They suffice to crack the atom, to command the tides, but they do not suffice for the oldest task in human history, to live on a piece of land without spoiling it.

—Aldo Leopold
Green Fire

I think it's time we talk about technology.

The plow was the obligate technology that dictated the cultural shift toward a belief in man's dominion over nature. Adopting a tech that is obligate across cultures changes behavior. And when behavior changes in large groups, societal values can change. And changing values at scale changes cultures. We saw how that played out with the plow. But what about all the technology we have now?

The internet, social media, AI,* quantum computing: these are obligate technologies. Any country that didn't or doesn't adopt them will fall behind as massive gains in productivity (and therefore wealth

---

* This book is already too long to talk about the threat of AI. But, you know, it's a real thing to add to your bingo card.

and power) are seen for the countries who did. But as we look at the tech that's become obligate both for nations and for individuals (it's very difficult to live a functioning life without internet these days), we have to ask ourselves: what values is our technology dictating? Individualism, success, wealth? And what values do we want it to dictate?

Technology over time has grown exponentially. We went from the railroad to the car to the fucking moon. We went from computers taking up entire warehouses to fitting in our pockets—with millions of times the computing power that launched us into space in the first place. According to Moore's Law, technology tends to double in capacity every eighteen months. That means it's cheaper and more accessible to more people. Awesome. But instead of using this incredible power of technology to make our lives better, we took the magic of the internet and used it to make more dumbass shit you can buy like avatars in the Metaverse or nonfungible tokens (NFTs) that are literally just pixels. They don't even exist in the world, and they're selling for millions of dollars. What the actual fuck?

But just because we've been using our incredible technology to make some of the dumbest shit imaginable doesn't mean we have to throw the baby out with the bathwater. There aren't only two choices: live in a cabin in the woods like Thoreau or have a smartphone but society is a dystopian wasteland. We are incredibly innovative. And we can make the tools do whatever we want them to do.

What if we created technology that instead of working solely to make more money for the creators of that technology, worked to make us happier? What if we could make technology that actively benefited humanity instead of doing exactly the opposite . . . technology that could guide us toward the change in values we need? Would we need to create new tools to do it? Or do you think we can maybe use some of the ones we already have?

A hammer can be used to build a house—or it can be used to murder someone. This is a value-neutral tool. Its value (whether good or evil) is determined by the person who holds it and what they do with it.

We all know that social media has been tearing apart the fabric of society with its incentivization towards polarization and vitriol. But social media in itself isn't evil. In fact, it's the most powerful tool we have to shift cultural attitudes and amplify positive messages. The technology that has become utterly ubiquitous in our lives—to the point where we can't imagine a life without it—is a value-neutral tool. I mean, it's not right now. Right now, it's a murder hammer. But it can be.

Social media algorithms could be maximizing for community, for wellbeing, for sharing, for happiness. They don't have to be designed to get us to spend as much time on them as possible. They don't have to be designed so that we're as angry as possible because that gets the most likes and retweets and keeps you on the platform longer. They don't have to be designed to radicalize us. The only reason they are optimized for this is because their business models are optimized for profit. But what if we incentivized them to do the opposite? What if we gave social media companies massive tax breaks if they could prove their algorithms were improving wellbeing or reducing polarization or teen suicide? Or even crazier, what if we just banned them from operating in a manner that's detrimental to society in the first place?

It's not that crazy. We already do it for lots of industries. We tax cigarettes and sugary drinks. We banned cocaine and heroin. We make it expensive for people to do things that hurt themselves, but we also make it expensive (or illegal) for companies to do things that hurt people. As soon as any industry has the power to materially impact nationwide social wellbeing, it should be heavily regulated—if not fully socialized. We regulate healthcare and the food industry because if we don't, they'll hurt people. We outlaw drugs that hurt people and lawn darts that hurt people.[†] But now the technology we have is hurting people, and we're not doing a damn thing about it.

[†] But never guns, of course. Guns don't hurt people.

What if any social media company with over a certain number of daily active users had to publish their algorithms and those algorithms were legally required to optimize for societal wellbeing? At the very least, we could stop it from turning our parents into QAnon believers.

If we don't do anything about it, social media will continue to shape who we are and what we want without us being aware of it. We are shockingly manipulatable. And they know it. And so long as they have that information, they can keep getting us to buy stuff we don't want, and feel shitty about ourselves, and then spend hours on apps designed to make us feel shittier. They're not doing it because they're evil. It's the world we've created; they're just maximizing for the thing we've all been told to maximize for. The only reason we haven't forced them to change the way they operate is because they have so much money and power and influence. But just because we haven't doesn't mean we can't. The EU just passed a law requiring social media companies to allow people to opt out of algorithm-based content and go back to the good 'ol "chronological timeline" which means no more "suggested posts" on Facebook or "suggested videos" leading you down the conspiracy theory rabbit hole when you're watching YouTube. The EU is also in a battle with Facebook over a ban on targeted advertising that harvests troves of users' personal data to show them highly targeted (and thus more effective) ads. Right now, it looks like Facebook is winning.

With every piece of technology that we've adopted since the invention of "infinite scroll," we need to ask ourselves: what is this maximizing for? What values is this propagating and instilling in us? And then after that, we need to ask ourselves: Is this even necessary? And finally: How much energy is it using?

Does anyone need an LED screen on their fridge that tells you what's inside the fridge? Lol, no. Does anyone need a smartwatch to let you know that you got a text so you don't have to check your phone that's sitting five feet away? No. Does anyone need a 3D printer? You bet your ass, we do! 3D printers save waste, democratize design, spur creativity, and help to re-localize production. That being said, *you* still probably don't need a 3D printer in your living room.

This is the goal as we think about all the technology in our lives:

how can we fuse the magic of modern human innovation with a simpler, happier future?

In *The Future is Degrowth*, they call this convivial technology. These are technologies that can help us work toward our goals rather than making it harder to achieve them. Convivial technologies don't have to cost thousands of dollars. They are technologies that are accessible to more people that make life better for more people.

And the fun part is, people are already designing and building all kinds of convivial technologies. People are making apps to give away free stuff to other people in your neighborhood rather than buy something new or throw something away, or apps to share compost in a city where most people don't have gardens, or apps to tell you the ecological footprint of a product before you buy it.

And technology isn't just an app on your phone or a machine that prints houses. It's any innovation that can change the way we live. And there are lots of ideas already out there to start changing the way we live and eat and connect and communicate for the better.

Convivial technologies include things like shared laundry in your apartment building or shared "tool lending libraries" in your neighborhood. Everybody doesn't need to own an impact drill or a 3D printer. But when you need to use one for a project, you can borrow it for the day. And the more tiny versions of these ideas start popping up, the better we'll understand how to scale them in society as a whole. People are already doing this all over the world.

There is already a movement for "15-minute communities" that localize more business and make it possible to exist without driving twenty minutes to get anywhere. During Cape Town's water crisis, they started a website to publicly post water usage in an effort to shame people into using less. What if your carbon footprint was public? What if we had individual carbon targets and everyone could see everyone else's usage? Would it help you to think twice before buying a new shirt you don't need? Would buying second-hand clothes become a whole lot cooler? Just as Adam Smith knew, shame works too!‡

---

‡   To be clear, I'm not attempting to put the burden solely on the individual. But shame doesn't work on corporations. We need regulations.

So how do we create and design technology in harmony with the biosphere? How do we reset our mechanical minds and our polarized minds and our consumerist minds while still harnessing the incredible power of those minds? But instead of using them to make money, we use them to make good?

If the biggest problems ahead of us lie in finding a balance with nature so we don't decimate the only thing that makes the food that lets us keep living, then let's start by applying our technological prowess there. If places are running out of water, how do we start thinking about more intelligent water management? If we're killing the land that grows our food, how do we un-kill it?

The permaculture movement provides a framework to think about this. It's not about being "natural" and letting the water go where it goes. It's about designing the landscape you live in to retain as much water as possible. Permaculture isn't some crazy hippie shit. It's people innovating in ways that understand how to work with the land and make the land work the best for us. Work smarter, not harder. And it's not like we need a bunch of new inventions, either. We already know a lot of ways to manage water more wisely. There is no reason the developed world should be flushing toilets with perfectly clean water when greywater flushing systems already exist. There is no reason we should be monoculture farming and stuffing the land with pesticides and synthetic fertilizer when we *already* know it's killing the soil, and we *already* know ways to make it better. This isn't solving nuclear fusion; we just have to start doing it.

Eight billion people can't live on the planet the way we're living now. But there is so much we could do literally today to fix the damage we've done.

The UN's Global Land Outlook report[112] estimated that five billion hectares—an area five times the size of China—could be restored by 2050 through basic changes in agricultural practices, including avoiding heavy tillage (fuck the plow), integrating trees with crops and livestock, and rehabilitating grasslands and forests. And obviously eating less meat that uses up an insane amount of land would help a whole lot, too. But many of these fixes are low-tech, accessible, and don't even require massive amounts of capital. The authors of that report estimate it will cost just $300 billion each year to achieve "significant" land restoration by 2030, which is far less than the subsidies already provided to farmers in developed countries. As a reminder, that's just a third of the United States' military budget—to restore 5 billion hectares around the world. If you recall, the entire biologically productive amount of land and water on the entire planet is 12.2 billion hectares. That sure sounds like a good investment to me.

And you know what happens when soil is more complex and has more life and larger root systems? It stores more carbon. A lot more carbon.

A 2009 report by the Wentworth Group[113] concluded, "At a global scale, a fifteen percent increase in the world's terrestrial carbon stock [including forests, woodlands, swamps, grasslands, farmland, and soil] would remove the equivalent of all the carbon pollution emitted from fossil fuels since the beginning of the industrial revolution."

Holy shit, are you serious? How are we not already doing this!? Can we all just go outside and plant five trees this weekend?

When we start to look at repairing our broken systems, we have to remember that science on its own (at least our current understanding of it) is woefully inadequate at explaining infinitely complex microscopic systems that co-evolved over millions of years. We have to finally admit we may not be as smart as we think we are. We have to break out of the mechanical mind and see that we are not bigger or better or the masters of anything. We have to understand that decimating a complex ecosystem to replace it with a monoculture crop that you pump full of fertilizer to keep it artificially producing food lacking most of the micronutrients we need is not a long-term solution. We have to go back to working *with* nature because that's literally the only way things work.

And we can do all this by harnessing and developing technology. We don't have to be Luddites. We shouldn't have a problem with

GMO crops. We should have a problem with GMOs where the seeds have been specifically designed so the plants don't produce viable seeds, so you have to buy more seeds every year, and they're only resistant to a single brand of pesticide, and a single company owns patents to 90 percent of the soy and 80 percent of the varieties of corn we grow, so it's illegal to save the seeds even if you want to! Can someone explain to me why we ever let a corporation patent the food we depend on for survival?

There are plenty of environmental nutjobs out there saying "only natural things are good." But that's crazy. Cancer is natural. Plenty of poisons are natural. Earthquakes are natural. What we should be looking for is technology that works in balance with the systems as we understand them—since our understanding of the systems has evolved since we first thought we knew what we were doing. Once we strip out the monopolistic control over industries that are critical to human wellbeing, we will see a lot more innovation that's geared towards making things better.

And the best part is, some of these technologies will actually make people more money. We don't all have to make altruistic sacrifices to make the world a better place. A lot of those regenerative agriculture techniques *in*crease profit margins for farmers. Farmers stop spending thousands and thousands of dollars on fertilizer and pesticides and rebuild the complex ecosystems that sort of take care of themselves. The land can hold more water, is more drought resistant, is more pest resistant, and supports more animal grazing when you do it right. On top of all that, they see higher crop yields and bigger profit margins than monocultured pastures drowning in synthetic fertilizer. It turns out cow shit is already fertilizer! And it's way better than the one we make out of natural gas! According to Charles Massy:

> In a healthy soil there can be up to 5760 kg of living soil biological community biomass per acre all busily teeming with life. Conversely, in a dysfunctional soil, there may only be 445 kilograms of bugs and creatures. This is the equivalent above the ground of between eleven yearling steers per acre on the one hand and less than one steer on the other. That is, one farm could be

fourteen times more productive in the engine room that matters. This in turn can result in around an extra one million litres of water in the soil bank of healthier soils.

Fourteen times more productive! You can have even *more* cows! And you know who eats all those bugs and creatures living in this gloriously rich soil? Birds, lizards, frogs, spiders! And you know who eats birds and lizards and frogs and spiders? Squirrels, foxes, snakes, cats, bats, coyotes . . . It's that whole circle of life thing!

The only reason more farmers aren't already doing this is because of late-stage capitalism. The agriculture industry is run by Big Ag, and they have a vested interest in keeping you addicted to their fertilizer and pesticides. Otherwise, their whole business model falls apart. When massive industries get threatened, they spend a whole bunch of money silencing the solution that would hurt their business, no matter how beneficial that solution could be for society. It's true of the sugar lobbies and the tobacco lobbies and Big Oil and Big Pharma and Big Ag and literally every industry because every industry we have is run by a monopoly. And when those monopolies with bottomless pockets get threatened, they pay for reports from seemingly academic institutes to tell everyone how wrong the new idea is. They conduct fake experiments to show how the way that keeps you buying their product is the only way. Smoking doesn't kill; nine out of ten doctors recommend Camels! High fructose corn syrup is good for you! It's those eggs and avocados you need to watch out for!

It doesn't have to be this way.

Industries don't have to have boundless control to write the narrative about what's best for society. We don't have to give them any control at all. Monsanto could only patent all the corn we eat because we gave them the power to do that. We can take that power away. It's worth remembering that none of these corporations exist without us supporting them.

The beauty of technology is its inherently democratic nature. If you're a farmer, you can change the way you farm today. If you run a small business, you can shift your supply chains today. If you're an individual, you can change the way you eat and live and shop today. Information is already out there. Technology is already out there.

If we start prioritizing the way we interact with nature and each other—and prioritizing our own wellbeing and future survival over short-term profits—we can work within our existing systems to bring about change. People do care. People are working hard to solve these problems. It may feel nearly impossible to make any progress in a system controlled by vested interests who have utterly lost the plot, but making the changes we *can* make is the first step toward changing the systems themselves.

Fuck it, dude. Let's go bowling.

**We must either learn to live together as brothers
or we are all going to perish together as fools.**

—Martin Luther King Jr.

Technology changes behavior that changes society that changes values. But if we're going to change our values, it might help to understand where they come from in the first place. Societal values aren't inherent; they're learned. And if they're learned, they're taught. And they change all the time.

Cultural materialism dictates a lot of this. It determines the systems we were born into that determine what we value and how we interact with each other. Technology is a piece of that puzzle. How we survive and eat and the weather where we live are all other pieces as well.

We think we exist as individuals, that our decisions are our own instead of being shaped by a greater societal force, but that's never been true. Everything we think and feel and the lens through which we see ourselves and the world is crafted and shaped by the society in which we live. We aren't born with any of it. An American baby adopted by Chinese parents and raised in China will become an adult with Chinese values. A Chinese baby raised in America by American parents will become an adult with American values.

When we start to look at these forces in our modern system, in our countries and towns and even neighborhoods, we can start to see how they shape us. And once we can step back far enough to see how they're shaping us, we can choose to start shaping them in the direction we want to go.

Back when Adam Smith was talking about using communities for social good, society was defined by conformity. "Fitting in" was valued, so it made sense people would make choices that would help avoid being ostracized from the rest of the community. But when John Stuart Mill rolled in, everybody started talking about wanting to be different. "Spontaneity, originality, choice, diversity, desire, impulse, peculiarity, and even eccentricity" were the values *du jour*. The new liberalism looked to define value and worth as being the least like anybody else.

For the next 150 years, since Mill's wild age of freedom and self-discovery, the pendulum of society has continued to swing. We've found ourselves re-constrained and un-constrained again and again. The Victorian Era was stuffy and morally repressed; Emerson and the Romantics sought to break free from that. The post-World-War-II era was defined by solidarity and thus conformity;[*] the hippie liberation swung back again. And those free-loving hippies were countered by the staunch Reagan conservatives who were countered again by (or possibly evolved into?) the frighteningly individualistic culture we're in now.[†] The point is it changes all the time. It's never *not* changing. And maybe we're right at the point where the pendulum is about to swing back the other way.

---

[*] This is why almost every woman born in the '40s and '50s was named "Mary." People craved unity, comfort, and belonging.

[†] This explains why almost every parent today is trying to name their baby some insane shit like Jaxsyn or Bayleigh.

I'm all for individualism when it comes to breaking out of the boxes society has told us we need to neatly pack ourselves within—when it comes to dressing differently and saying how you feel and quitting a job you hate to follow your passion—but our fierce individualism becomes a problem when it threatens the social fabric within which we all exist. As Americans, we believe in the sanctity of the self above everything else, no matter how many other people get trampled along the way. Just think for a moment of all the people who refused to wear a piece of cloth on their faces to save other people's lives because it "impeded upon their personal liberty."

Call me radical, but the whole 'individual liberty' argument falls apart when it starts killing people. It falls apart when we refuse to regulate the guns that continue to kill hundreds and hundreds of children (and adults, and everybody) because my right to own one is more important than your right not to die from one. It falls apart when the way we're living is destroying the society we depend on to live.

We all have our constitutional right to life, liberty, and the pursuit of happiness, but maybe we've taken it just a little too far. Liberty is the freedom to do your own thing, sure. But it's not in the absence of any social constraint or legal constraint or moral constraint. We're not all supposed to just follow our passions with "unbounded will and unconstrained desire."[114] It's not okay to lie, cheat, or steal or do whatever you have to do in order to fulfill this ideal individual self.

"Liberty" isn't indulging every whim that strikes your fancy and believing that any attempt for the government to make laws or help regulate society is a form of tyranny. The neoliberals have gotten it all wrong.

Liberty is living in a society that's built so that we can all be happy. It's a society that provides the basic necessities to people so that we can all live our best lives. Liberty is sending your child to school knowing they won't get murdered there. It's the freedom of knowing that even if you lose your job, you'll still have healthcare. It's the freedom to quit a job you hate because the system is robust enough to support everyone. But, as the adage goes, freedom isn't free. There is still something that has to be given up. Well, it might seem like you're giving up something, but it doesn't have to *feel* that way. Maybe we can transition to something that's both—that's both community-based but also based in self-interest. If our most basic desire is to be happy, maybe we just have to be willing to reframe how we see that, how we define it, and how we go about getting it.

While many Asian cultures are strictly collectively oriented—the individual always comes in second to societal identity and need—it's not all or nothing. You can have your cake and eat it too. Scandinavian cultures are both personally individualistic and communally collectivist. Sure, you can pursue happiness in whatever way you see fit. But they also understand that certain community values are simultaneously better for the individual.

Working together and everyone helping each other is better for you and better for me. If I have to do everything on my own, life can be pretty hard. Maybe Scandinavian societies evolving in a brutal, arctic climate where the sun barely rises for half the year actually had some benefits. No one would have survived if they weren't willing to help each other out.

It's the selfish case for being ethical: any kindness is always good for everyone—including yourself.

Here in America, we've been told that being kind means we're some kind of pushover, that we need to be tough in this tough world full of tough people. But being kind doesn't mean being weak.

What does a Buddhist do if someone steals their purse? Compassionately hits them with their umbrella.

Being kind means standing up for the values that make everyone better, and that makes things better for everyone. You know, like *not* stealing things or murdering people. It means caring when someone else is being abused or marginalized and standing up for them too. And that is a critical piece of every society. Because the more selfish everyone acts, the worse society is for everyone—including yourself. That's the rub. Even if you think acting in your own best interest is helping you, it's not. It's actively making it worse.

The problem is we've all been conditioned to believe this is the way we have to be. We think if we're nice or honest that we're at a disadvantage to all these cut-throat actors out there, but they are the few, not the many. And the worst part is, it's a self-fulfilling prophecy. People capable of and wired for kindness feel they have to be tough and selfish in a dog-eat-eat world where everyone is looking out for themselves. So they harden themselves, which hardens others, which brings the prophecy to fruition.

It's just a Catch-22, another multi-polar trap. Most people *are* kind. We've just been told being selfish is the only way to survive. We've been told when the zombie apocalypse comes, everyone's going to immediately start murdering one another, so your only choice is to start the murdering first. But it's not true unless you believe it. Unless you make it true.

The good thing is, we don't all have to sit down at the same time. You can stop being an asshole right fucking now. It will make you happier right fucking now. Like literally today. The moment you are kind to another person, you feel it. The moment someone answers you with kindness, you feel it. And answering your own hurt with hurt never helps.

It's not the dessert we ordered, but it still hits the spot.

Adam Smith saw community-based living as a lever for social good. He saw it as a method of reining in our baser tendencies and desires. But what if community is a whole lot more than that? Smith never contemplated a world without community because that simply wasn't an option back then. But now that we've seen what it looks like when everyone is isolated, alienated, and disconnected, we can see the power of communities goes far beyond keeping us morally constrained.

When we go back to integrating with our communities—with other people—we can be reminded of what's so great about being human in the first place. If the pandemic taught us anything, it's that being alone is bad for humans. We don't just want connection, we need it. And the more of it we have, the better. Old people who are lonely are 30 percent more likely to die sooner than old people who still have love and support in their lives.[115]

There's just one catch. Teaching our children (and ourselves) to be kind and empathetic and caring and giving is something we have to *do*. It's something we have to choose to foster. Here's Peter Whybrow:

> Selfish behaviors are reward driven and innate, wired deeply into the survival mechanisms of the primitive brain, and when consistently reinforced, they will run away to greed, with its associated craving for money, food, or power. On the other hand, the self-restraint and the empathy for others that are so important in fostering physical and mental health are learned behaviors—largely functions of the new human cortex and thus culturally dependent. These social behaviors are fragile and learned by imitation, much as we learn language. To be sustained across generations they must be carefully nurtured by extended families and viable communities.

The only way to get that community and love and support is to build it. To teach it. To actively go out and get it. If that's the piece we have to "sacrifice" for our own happiness, then that sure sounds worth it to me.

But what does that look like, exactly?

Well, first, we need to refocus our economies toward local businesses selling goods to local people. The shift from Main Street to shopping malls to big box stores to Amazon has razed the connection that local commerce provided between neighbors and the connection that meaningful work has to our own self-worth. And it's meant a higher energy throughput for basically every product we buy.

Sitting all day at the aglet factory running the machine that makes one tiny part of the aglet makes us feel . . . shitty. It makes us feel lost and worthless. Humans need to feel that the work we do has value. It's why so many people are so unfulfilled in so many bullshit jobs. It's why so many people are opening Etsy stores for their hand-knit scarves and hoping they make enough money to quit those shitty day jobs.

At the risk of sounding like a broken record, massive multinational corporations are destructive for society in basically every way. They took our civil society—the one based on a stable human community of collaboration and mutual respect—and replaced it with ruthless competition. Sure, things got cheaper. But cooperation, creativity, artistic expression, craftsmanship: those things are all devalued when you can buy that scarf that was made in Malaysia by little slave children for two dollars. And of course you don't know it was made by little slave children because we're not at all connected to the communities where the products we use are made. And we can't afford to live devoting our lives doing meaningful work crafting things, so we're forced to get a job at Walmart selling the two-dollar scarves and knitting our own scarves as a hobby on the weekend if we're lucky.

If the worst part of capitalism according to Marx is the alienation that comes with making a tiny piece of an aglet that disconnects you from making the shoe, we need to go back to making the whole damn shoe. We need to engage in work that is non-alienated, socially meaningful, self-determined, and dignified. Make those shoes, make those scarves, and sell them in a community full of people you know. Monopolistic capitalism shuttered local businesses because people will always buy the cheapest, easiest thing in the system we've got. Especially when you're struggling to get by and can't afford to make a more ethical choice. But when the globalization that supports slave-labor products being made for pennies collapses, we're not gonna have the luxury of a choice.

And once again, the shift is already happening. The refocusing toward highly skilled, artisanal products is an organic one. People want those locally crafted, handmade goods. People are already willing to pay more for these things. Just look at Etsy. People are already aware—even if it's on a subconscious level—that this is a better way for the world to be. But we most often vote with our wallets, and so long as we allow a system to exist that exploits human labor for corporate profit, too many people will keep buying the cheapest option no matter how many slave children made it or how many oceans it crossed getting made. It's just another case of by design or disaster: we can choose to start putting our money where our mouth is, or the collapse of the system will force us to before we have the structures in place to soften the blow. And the more local our products are—the closer we are to those means of production both physically and mentally and geographically—the smaller our energy throughput gets and the happier we get.

Don't worry. You can still sell your scarves on the internet; we don't have to close down Etsy. But we need that community so much more than we realize. And even though, yes, scarves will be more expensive, the price we're already paying is far greater than any monetary cost could ever be.

When we start refocusing commerce on communities, we'll start reconnecting with the people in our communities. We don't just need to reconnect with the means of production; we need to reconnect with the rest of the real live humans in our lives. We need to bring back bowling. Bring back knowing your neighbors and hitting the lanes once a week or starting a book club or a Rotary Club or invent a whole new club. If we can get to the world that late-stage capitalism was supposed to provide, the one where our basic needs are met and we're all working fifteen hours a week, we'll have time to spend forging and nurturing these relationships that are so vital to our ability to thrive. And the more time we spend in those communities, the better we become.

We think it's so easy to be selfish and isolated because we never have to deal with the hard work that often comes with human relationships that are endlessly complex. But the part we so easily forget is that the work it takes to build and nurture friendships has a nearly infinite ROI. It's the thing that makes us the happiest in the entire world. Think about your spouse, your best friend, the people that mean the most to you in your life. Now think about living in a world even more full of wonderfully fulfilling relationships. Relationships not only make us better people, they are an endlessly renewable source of lasting, deep, meaningful joy in our lives.

I'm not gonna lie; it's not gonna be easy. We're so hopelessly polarized and pitted against one another every day. We're utterly bombarded with this messaging from every media outlet in our lives. But the irony lies in the fact that all of this extreme tribalism is borne from our longing for connection and relationship. We evolved in our tribes of hunter-gatherers, programmed to participate in community. The tribes with the strongest community bonds are the ones who would survive. We are better together.

We just accidentally ended up with the most toxic, shallow form of connection you can possibly imagine. We created a society so severely lacking in community and belonging that we're willing to take it anywhere we can get it. We feel like we're part of a group when we call someone a libtard or a snowflake, or rail against anti-vaxxers, or "cancel" some actor who died six decades ago, but it's all just a placeholder for the community we're craving. People don't become flat earthers because they really believe the Earth is flat. They do it because they're so desperate for belonging, they're willing to believe anything. We have to stop seeing people as "others" and realize that we're all human. We need to refocus the locus back on the communities that make us better. Religion used to be the tie that binds us, but with a shift away from organized religion and away from local communities, there is a void we're desperate to fill. We can try to fill it with stuff and more stuff and vitriol and polarized tribalism, or we can fill it with empathy and love and real human relationships.

If we move the center of labor back to communities, we'll shake off the hangover of isolationism and remember how fucking awesome it is to have friends and be a human. And then Adam Smith could stop rolling over in his grave because his whole idea works again. Humans want communities, our system needs communities, communities create the virtues we hold dearest like love and trust and empathy and responsibility. So how did we let ourselves get so far away?

Material Girl in a Material World

**It's easier to imagine the end of the world than
it is to imagine the end of capitalism.**

—Fredric Jameson

We have a system that maximizes for profits because that's been the best way to continue growing for the past couple hundred years. It's been the best way to get the most for the most people. Sort of. But what happens when maximizing for wealth creates the problems? If wealth is connected to surplus—who has the most wheat—what happens when destroying the planet no longer gets you more wheat? What happens when the next obligate technology or obligate ideology is exactly the opposite: the cultures and nations who survive are going to be the ones who figure out a way to exist with and within the ecosystems they depend on (provided the self-interested chickens don't peck everybody's eyes out first).

Cultural materialism tells us when we change the mode of production—the way we get our food, the way we survive—society changes. Culture changes. If we shifted to a hybrid capitalist system supported by universal basic income but with guardrails against profits at the expense of human wellbeing, what kind of changes would we see?

If we democratized more of the means of production and basic infrastructure, such as electricity, transportation, and communications, how would it change people's behavior?

If we were free to pursue things we were passionate about instead of slaving away for a boss who doesn't give a shit about us (or only slaving away for fifteen hours a week), how would that change how we interact with each other? If no one was worried about where their next meal came from, would we treat each other differently?

The feeling of discontent I shared at the beginning of this book isn't just a coincidence. It's a side effect of the systems we've been told are the only way anything can ever be. So are authoritarianism and nationalism and everything else that's further dividing us. Nazis aren't making a comeback because humans are inherently or secretly racist. They're just a downstream effect of the runaway train of wealth inequality we've allowed to keep speeding up for the past four decades.

We're being torn apart by the artificial scarcity created in the system we invented. We're scared for our futures and scared for ourselves, and no one feels like there's a damn thing they can do about it. We've been told the pie is only so big and every brown person who crosses the border means a smaller slice for you. But that's not true. That's never been true. If we've learned anything from *Superabundance* it's that there are more and more resources to go around for everyone!

Scarcity is imposed by a capitalist means of production as a means of driving profit. The scarcer something is, the more they can charge. The scarcer jobs are, the less they can pay you. Diamonds aren't rare at all; DeBeers just keeps them all in a big safe and strategically releases them to hold their value.

There is plenty for everyone to live right now (though admittedly not in the ways we're currently living). We have the capacity for a world of sufficiency. A world that satisfies the material needs for humans globally and within planetary boundaries. There is natural scarcity we need to worry about, but there is plenty of public abundance to go around if we just change the rules of the game. And when we change the rules of the game, the way we treat each other changes. Our behaviors and values change.

We have to break our minds out of the confines of late-stage capitalism as the only system we've ever known and be willing to imagine something different and better. Part of the point of this book was to remind you just how new all these systems and ideas are. This isn't the only way things have ever been or the only way things can ever work. This is our very first attempt at capitalism. So we figured out a few things that capitalism is good at (breeding innovation and creating wealth) and a few that don't work that well (massive inequality and debt and unsustainable endless growth). If I know anything about how anything works, it's that nothing is black and white. Everything in this world is nuanced. Unbridled free-market capitalism isn't the answer any more than communism or fascism is. Everything is always complex. Always. And the answer we're looking for is surely some mix of capitalism, and socialism, and degrowth, and toss in a little animism and environmentalism and hedonism and every -ism you can think of. Except Nazism. We don't need any more of that one in there.

The point is we can't yet imagine what this future system is going to look like because it's never existed before.

Maybe the depletion of finite resources will precipitate massive changes in the way we interact with each other for the better. Maybe we'll figure out some sort of solidarity economy that values cooperation before competition and purpose before profit. I mean, that's probably not going to happen until most of the world is destroyed, but it could! And it could also happen if we just make it happen before everything collapses—by design or disaster, amirite?

If you want to get even more kumbaya, we can start looking to an economy built out of cooperatives, confederations of citizen initiatives, and local democratic communing practices. It's already happening. According to *The Future is Degrowth*:

> In Catalonia, Spain, a cooperative of 2,500 members runs exchange networks, its own currency, food pantries, assemblies,

financial cooperatives, a collectively-run factory, a machine working shop, and supports around forty-five people with a basic income. The Catalan Integral Cooperative, founded in 2010, is an amorphous network whose main mission is to 'antagonise Capital by building cooperative structures in the Catalan economy'. Since its foundation, it has developed several diverse, but interdependent, initiatives which have as their explicit goal to displace the state apparatus—covering health, food, education, housing, and transport.

The cooperative has become an encompassing network that allows many to move much of their life outside of the dominant economic system. This involves, for example, participating in one of the many committees which decide the direction (legal, financial, technological, and so on) of the network. Involvement in the work of the committee also implies receiving a basic income, partly in euros and partly in their own currency system. There is also a well-developed local exchange network, which supports autonomous, small-scale production as well as 'pantries' which are connected through an internal transportation and logistics system. The cooperative also includes many autonomous organizations, such as events spaces, cooperative housing units, and the impressive Calafou—a 'postcapitalist ecoindustrial colony' in the ruins of an abandoned industrial village in the Catalonian countryside. In 2017, Calafou was inhabited by two dozen people and, on top of that, housed a carpentry and mechanical workshop, a community kitchen, a biolab, a hack lab, a soap production facility, a music studio, a guest-house, a social centre, and a 'free shop'—each run collectively and non-hierarchically.

By itself, Calafou is certainly unique. But what makes it so special is its connection to an expanding ecosystem of other similar projects through the cooperative and its many members. This innovative project is emblematic of what Wright calls an 'interstitial strategy', as it allows its members to experiment with different ways of organizing housing, food supply, technology, currencies, and the revaluation of labour—away from an exploitative, alienating system towards one that is needs-oriented and meaning-making.[116]

That's fucking awesome. Where do I sign up? And there are a million other ideas and movements that are already happening as well. Some of them involve trying to work within the system we have, but plenty involve working around it, outside of it, or simply existing within the cracks. Radical municipalism recognizes that voting at the ballot box for candidates who support a $15 minimum wage or paid family leave or platitudes about social justice is never going to be enough. We have to start rebuilding our systems of governance from the bottom up, starting with town councils and local assemblies.

Libertarian socialism calls for the complete destruction of the authoritarian institutions we have that control the means of production and that subordinate the majority of us to the economic elite. You wanna get even crazier? Get involved with social anarchism, which doesn't just want to dismantle the system, but wants to leave almost nothing in its place. In this radical future, we would govern ourselves solely through a network of voluntary associations and local committees.*

Another real wacky idea for local governing is known as watershed democracy. In this system, political boundaries are redrawn based on the watershed that feeds an area. Apparently, villages in India have been living this way for millennia. Which makes sense. Because your county or state should be making decisions that best manage the water that feeds into your land. But when I live upstream from you, I can just take as much as I want without worrying how much anybody else gets. When we sell water rights to Nestle, they have no reason to do anything with that water other than suck the land dry, put it in a bottle, and sell it right back to you.

But while we need to invent a new system from within the society we accidentally created, this isn't about imagining a new future from scratch. Plenty of versions of more equitable, localized democracy have

---

* This one might be too liberal, even for me.

already existed in humanity's past. We just erased them from the history books because they didn't fit with our narrative of "white Europeans are civilized and intelligent and everyone else is a goddamn savage." Do you know how crazy Native Americans thought European aristocracy was? They literally wrote essays about it. In *The Dawn of Everything*, Davids Graeber and Wengrow give numerous examples of equitable, organized societies that provided housing to everyone, were ruled by large citizen committees, and even societies that consciously reversed inequality to improve the lives of their people and avoid collapse. Humans are smart. We can do this. We already did this. And it was working just fine until those musket-wielding capitalists showed up.

The point is that anything is possible. And people starting to change the way they live on an individual and municipal level will change the way we live on a societal level. As author, journalist, and radical municipalist Debbie Bookchin writes:

> The great news is that this politics is being articulated more and more vocally in horizontalist movements around the world. In the factory recuperation politics of Argentina, in the water wars of Bolivia, in the neighborhood councils that have arisen in Italy, where the government was useless in assisting municipalities after severe flooding, over and over we see people organizing at the local level to take power, indeed to build a *counterpower* that increasingly challenges the power and authority of the nation state. These movements are taking the idea of democracy and expressing it to its fullest potential, creating a politics that meets human needs, that fosters sharing and cooperation, mutual aid and solidarity, and that recognizes that women must play a leadership role.[117]

We don't all have to go join self-sustaining hippie communes, but doesn't it sound kind of . . . better than your shitty job?

There are plenty of people out there who will tell you I'm crazy just like Malthus. They will dismiss this book and these ideas just like they've been dismissing everyone talking about these things since the first time they were ever talked about. There are plenty of people who dismiss "apocalyptic environmentalism" as a pseudo-religion, as if that somehow negates the value of it. Michael Crichton, the author most famous for writing *Jurassic Park,* once noted:

> If you look carefully, you see that environmentalism is in fact a perfect 21st-century remapping of traditional Judeo-Christian beliefs and myths. There's an initial Eden, a paradise, a state of grace, and unity with nature, there's a fall from grace into a state of pollution as a result of eating from the tree of knowledge, and as a result of our actions there is a judgment day coming for us all. We are all energy sinners, doomed to die unless we seek salvation, which is now called sustainability. Sustainability is salvation in the church of the environment. Just as organic food is its communion, that pesticide-free wafer that the right people with the right beliefs, imbibe.[118]

I mean, he's not wrong. And that's fascinating. But if religion is a piece of cultural materialism, and the movement from animism to monotheism was precipitated by the domestication of animals, then maybe this is just the next step in our cultural evolution. Humans need to believe in something; we don't work any other way. Our brains got too big, and now we need to find meaning and purpose in our lives. And if the meaning in our lives that was once largely inhabited by animism, and then polytheism, and then by Christianity and Judaism and Islam is being replaced by a reinvigorated love of and appreciation for life-giving nature, I'm not sure I see the problem.

Christianity doesn't have a whole lot going for it right now. They lost that whole Jesus vibe, "love thy neighbor as thyself" plot a long time ago. I'm pretty sure Evangelical Christians are the most hateful people in America these days. And I gotta be honest, the Israeli Jews aren't doing much better. They've created a brand-spankin'-new Apartheid state #sorrynotsorry.[†] Islam? Yeah, again. Plenty of wonderful, kind, honest people in every religion. And plenty of jacked up movements making good people believe they're justified in doing bad things in the name of their dogma within each one as well. So I'm not really sure why anyone would be signing up. All organized religion ever did for me (and my father was a pastor) was show me how easy it was for sanctimonious people to be complete hypocrites. Except Buddhism. Those guys seem to know what's up.

Would it really be so bad if we all just forgot about the Crusades and the Inquisitions, and the holy wars and *jihads*, and just ... decided to start worshipping the sun again? It literally provides ALL LIFE ON THE PLANET. I can't think of a better thing to worship. If loving nature and finding meaning and purpose in living in balance with nature and the rest of the humans on this incredible planet is wrong, then I don't wanna be right.

I mean, have you ever been out in nature? Nature is fucking lit. There is nothing manmade in this entire world that can hold a candle to the Dolomites in Italy. Or the glacial lakes of Montana or the Lost Coast of California or the Canyonlands of Utah or all of fucking New Zealand. Or SCUBA diving in a multi-colored reef or the turquoise waters of the Caribbean or the flour-soft, pink-hued sand in Zanzibar, or the motherfucking Redwoods. I realize I am incredibly privileged to have traveled to all these places, but there is something beautiful to be found on this planet everywhere you look. Call me a tree-hugger if you want because I'm here for it.

I'm here for whatever changes help us rediscover what we've forgotten. I'm here for whatever changes come as a function of cultural materialism—whatever changes in our ideology come about because we've been forced to change the way we live in order to survive. I'm

---

† I wrote this long before the latest installment of the Israel-Hamas war began. But let's just be clear: everybody is in the wrong here.

here for any idea or shape or concept to restructure the world into a way that works. If nature is our new religion, and we go back to celebrating the spring equinox like pagans dancing naked in the forest with rituals of rebirth and renewal (which is why Easter is all eggs and flowers, btw), then where do I sign up?

Where Do We Start?

The grand essentials to happiness in this life are something to do, something to love, and something to hope for.

—George Washington Burnap

If you've come this far and you're ready to do the work, great. The problems are massive, they exist in every sphere of the lives we lead, and more than likely, shit is going to get a whole lot worse before it gets better. But there are things you can do besides eating fewer cows and turning off the A/C.

The first thing you need to do, according to Jamie Wheal, is get your house in order. Look at your life and figure out the things that don't fit with all the new information you're synthesizing into this new world view. Understand the changes you can make, the ones you can't, and be okay with what that looks like. I'm still struggling all the time with how much I fly. But I live in Europe, and my family is in America, and I love traveling, and it's hard. I've tried taking trains and buses, and even within Europe, it's shit. It's way more expensive and takes seven times as long. Sometimes I still take the three-hour flight over the twenty-hour bus. But I'm okay because I know I'm aware of it and working on it. I know what I'm working to do, what I'm working toward. My house is in order. The only way to suddenly live a completely sustainable life within the constructs of the society we have is to fuck off to the forest and shut yourself off from the rest of the world. But we can't fix society from outside of it. We have to find the best version of our own lives to lead within it.

There's a concept of right action that someone else came up with, and I have no idea who, but the idea is that whatever you decide your ideals are, whatever things you believe define the way you should live, you're never going to perfectly fit within those things. But there's a balance. Just like everything else in the world, it's nuanced.

If you create a set of ideals and rules you're supposed to live by, and you do ten or more things in your life that don't abide by that system of rules, you're a hypocrite. You can't say you care about the planet and want to effect change but then keep on living your life exactly the way you've been living it. You can't keep taking flights all the time, and eating McDonald's every day, and blasting the A/C in September, and working in your corporate job as a middle manager for an oil company whose main objective is to cut costs by firing people and slashing benefits . . . and then go home to your family and tell them how much you want things to change. You have to walk the walk to some degree. Otherwise, what's the point? But the other side of this is, if there are fewer than four things that break the rules, you're a fanatic. I saw an interview with a woman advocating against single-use plastics saying, "Well, I still keep Ziploc bags in my drawer, they're just so convenient!" You can work to change a thing while still being a tiny bit of a hypocrite about it, I promise. Hopefully that lady at least washes her Ziploc bags instead of throwing them in the trash.

Once you do start making changes, don't beat yourself up just because you're not perfect. We all still have to live in this society as it exists now while we work to change it. So do what you can do, and give yourself a break.

The next thing is to figure out exactly what it is that you can do. Figure out what you're good at, where your passion lies. The most valuable thing anyone can do is to devote their time, their work, their life toward fixing some part of it.

Can you simultaneously contribute and fulfill your heart's desire? If you could imagine the best day, doing exactly what you love doing, what would it look like? And I don't mean the first two weeks you'd spend doing nothing because you're insanely overworked and just need a break. I mean after that. If you could spend the whole day knitting or drawing or building things or fixing things or talking to people, or being inside or being outside, whatever it is, is there a job you can do that uses that passion?

And once you know what type of job that would be, start thinking about ways it might fit into the bigger puzzle. We're all so tired and overworked and change seems so scary, but it's not. It's exciting and life-altering. Haven't you ever dreamed of something other than the drudgery of your daily existence? Have you ever had a spark of an idea, and then immediately thought, "I can't do that, I can't afford to chase that dream." Well, you can. And maybe your idea was an amazingly altruistic one. Like my friend Natalie who wants to start a music therapy camp for refugees. But her immediate response to herself was, "That's a stupid idea, refugees don't need music." But maybe they do? Maybe they need healing a whole lot more than the rest of us. It doesn't solve climate change, but it sure as shit makes the world a better place. Spending time making music is time that's not spent buying more shit or watching TV or driving somewhere. Or what about my friend Maria who believes that helping people communicate better—with honesty and vulnerability, without fear of shame or ridicule—would help to solve most of the problems we're facing. Goddamn, she's got a point.

Or maybe you want to start a nonprofit that buys up land and then helps people become homesteaders? Or a media company that works to deradicalize people or stop fake news from bot farms. Or maybe you've got a convivial app idea or maybe you just want to knit your scarves all day because that should also be a viable means of existing on our planet, and everybody needs scarves. There are enough of us who want something different and who believe something better is possible, I promise. We don't all have to become eco-warriors or sustainable farmers. That's sort of the whole point. It wouldn't even work if we did.

In my case, I was inspired to write this book. For the past seven months, I've spent every spare moment I had writing this book because it's the one thing I could think to do right away that might help to solve the problem. And even if it doesn't solve anything and no one reads it, I still get the satisfaction of working toward a purpose I believe in. I have some other ideas for communal, sustainable living I'm working on as well. If you have a skill or a passion or an interest that's part of the solution, do it, follow it, write it, invent it, talk about it. Talk about it with your friends or start a podcast and talk about it with everyone. If you can find a way to make a living while being part of the solution, that's the whole idea.

For most of us, quitting our jobs today to knit scarves probably isn't an option. I get it. I still have to edit people's books and write advertising copy for large corporations to perpetuate their need for endless growth. But I do as little of that as possible to support the things that matter the most to me.

If you have to keep your bullshit job to eat and pay bills, think about the least amount of work you could do. How many hours of the work you do pays for your house and bills and car, and how many hours is extra? How many hours of work goes to pay for designer sunglasses and new clothes and other things you don't need that aren't making you any happier?

We each have to figure out our own baseline for living. First get rid of excess so you arrive at "comfortable." Hopefully you did that a few chapters back. And then see what you can take out of comfortable to get to "I can live with this." I could certainly do a better job of that. We all could. It's not easy in a system that doesn't yet have outlets to do it. And the Incredible Everything Machine sure is incredible. But I'm working on it.

Now the idea is to work the fewest number of hours to get you to this reduced level of consumption. There is a bougie millennial movement called "Coast FIRE" based on the Financial Independence Retire Early principle. FIRE is a mindset or life approach where people working their asses off at $200,000-a-year jobs spend as little money as possible so they can retire as early as possible. Can everyone do that? No. But the concept is still valuable: how little can I live on to give more hours, more years of freedom to myself? Coast FIRE is similar except you make enough money to just work part time, or coast, the rest of your life. But in our version, you spend the rest of that free time working toward the piece of this puzzle you care the most about.

I understand there are plenty of people absolutely struggling just to survive every day, and thinking about the climate or the future isn't really an option. But I challenge you to think outside that box for one second. I challenge you to think if there is any other job, any other lifestyle, any piece of your life you could point in the right direction.

And if you hate what you do and hate your job and know your industry is terrible and you'll never have enough time outside of your eighty-hour workweeks—you have to be ready to blow up your life. You have to be ready to take that leap and believe there is plenty of happiness to be had on the other side. And hardship too. But working toward a thing you love, a thing you believe in, is infinitely more valuable and rewarding—and provides increased wellbeing—compared to having money and financial security doing a thing you hate. I promise.*

And it's certainly better than doing a thing you hate that *still* doesn't provide financial security.

The most important part of these approaches to life and work is, once you have that free time, however you get whatever free time you get, start devoting it to whatever piece of this crazy complex puzzle scares you or inspires you the most.

Understanding how vast and widespread these issues are may feel overwhelming, but it can actually be a good thing. Because it means we all have a chance to make a difference in a different place. If everything needs fixing, then there's work for everyone to do. And if we all find some piece of the puzzle that motivates us or makes sense to us—and

---

*   If you don't believe you can quit a job you hate, please read my first book, *Authentic AF*.

if everybody's doing that—then suddenly there is a revolution. A revolution of thought and of action and of values. Suddenly it's top-down, bottom-up, and inside-out all at once.

Maybe it's the Amazonian hydrological pump or maybe it's sustainable farming or low-income housing or local politics. It can be literally anything. Because there are an infinite number of ways to be a part of the solution. Everyone has a talent they can give. Maybe you're not even sure what it is because you've been working in shitty, soul-sucking bullshit jobs your whole life. But you have a spark. Everyone does. Start thinking about it. Start discovering it. Start trying things. And then start spending your free time working toward the thing. Give whatever talent or idea you have toward the thing. Because everything is both top-down and bottom-up. It's changing laws and geopolitical structures, but it's also the movement of the people. It doesn't work unless we start living it and sharing it and talking about it and working to change it.

The best part is, people are moving into positions of power who have these beliefs. It's already happening. We just need more. Do what really fulfills your heart and uses your abilities. No one dreams of a bullshit job. No one is fulfilled by money. No one.

And if you can't leave your job, find ways within it that could be part of a solution. My husband teaches ancient history to seventh graders. He can't change the curriculum, but he did come up with the idea to offer an elective about where our stuff gets made and how it gets to us from all over the world. It's introducing kids to the insanely complex, nearly invisible global supply chain system that most of us are completely oblivious to. That's an important piece of the puzzle too.

What if all the aerospace engineers graduating from engineering school refused to design or build new private jets? What if they all just said, "No, I'm not going to be a part of this thing that I know is terrible." Then there couldn't be any more private jets because there would be no one to build them. Think of it like a union. Unions exist because one voice isn't loud enough to fight against corporate interests. But the more voices you have, the more power you have.

If even a quarter of the country starts thinking about change and starts changing their daily behavior, change will come. It's impossible not to.

People have been making a difference for all of time. People are the only thing that has ever made a difference, the only force that can overturn systems built to work against them. The belief that we can't change anything because our voices are too meek is the most powerful tool the ones in power have. And you better believe they know that.

Once you've got your house in order and you've figured out what you can do and start doing it, the final piece is to help others figure it out.

In my opinion, this one is the hardest. No one wants to hear these things. I can't believe you've gotten to the end of this book. As I've been writing this book and trying to talk about it with friends, I've been dismissed time and again. It's been heartbreaking. When I came back from that trip to Southeast Asia last summer, I was so fully energized and positive, and I truly believed everyone would be as immediately blown away and instantly changed as I was. I was quickly crestfallen. I recently walked away in tears from a dinner date with some of our closest friends when one of them said, "I'm never going to want to talk about this. So just stop." I came back to the table after a few minutes, attempting to gather myself though unable to stop the tears still streaming down my face, and ate my ramen in a heavy silence. I really am no fun at parties.

It occurs to me now that my last massive paradigm shift followed a strikingly similar pattern. It also followed a trip to Southeast Asia. It was the first time I realized how little it took to be happy. That I didn't need any of the clothes and cars and shoes and stuff I had worked so hard to accumulate. But when I returned from that trip, believing everyone would be receptive and similarly awakened by this epiphany, nobody cared. Nobody cared what an unemployed couchsurfer with greasy hair and holes in her sweatshirt had to say about anything. It will be the same for this, and it will be disheartening. There's no way around that. But sometimes it will be energizing. Sometimes you'll meet a person who gets it. Who sees how simple and complex the whole thing is, and when that clicks, it's all worth it. If you can help just one person figure it out, if you can change one person's mind about what's possible and what we need to do and what the future looks like, then everything you did was worth it.

And the more people believe there is an answer to the problem, the more agency is created in people to work to solve it. It will give people who already have the heart and the mind the will to start working toward it. Believing that nobody cares and everything is fucked is the biggest part of the problem. It's the only thing getting in the way of changing anything. But imagine if just ten percent of Americans—34 million people—were feeling the same way you were? Imagine if all the people feeling the same way, and feeling like it was pointless to try to fix anything, could come together? When you see other people walking a path that makes sense to you, it gives you the knowledge and courage to follow it. The more people who do it, the more people can.

I promise you, it's more than ten percent of us. We're all just drowning in this system with no lifesaver to grab hold of. But coming together is the lifesaver. Making small changes is the thing that makes the big changes. Changing what people believe is possible has to come before the change itself.

Hedonism is a Reasonable Response

I can't go on. I'll go on.

—Samuel Beckett
The Unnamable

I'm pretty sure that for most of us, the problem isn't that we don't see the problems or don't want to change anything. It's that it's so terrible, it's hard to fathom and even harder to sit with. I get it.

When you really start wrapping your head around all these problems, when you start to understand the vast nature of it all—how widespread, how deep, how systemic, how it touches on and is affected by literally every single aspect of the lives we lead—it's really fucking easy to fall into nihilism or hedonism or apocalyptism. Or a dash of each. When you finally let yourself arrive at the tragedy—if you let yourself arrive at the tragedy—it's fucking rough. It's so much. It's too much. When you realize how fucked everything really is, it's not uncommon to sink into a deep depression. You realize there's no saving us or the planet so why bother trying; who the fuck cares? My friend who made me cry at dinner isn't a climate denier or an asshole. He just cares so much and can't see anything but the tragedy. He cares so much that even talking about it breaks him. His only way to exist in the world is to simply pretend it isn't. Willful ignorance is his only defense.

For others, instead of ignoring it, they let the tragedy consume them. For some people who truly understand everything and truly care, proposing solutions is anathema. There's no point anyway because everything is totally and completely fucked. You're not taking this seriously enough! We can't change the world with a stupid little community garden. I don't see why I would change my life at all when everything is going to shit anyway and drying my clothes on a rack and biking to work isn't gonna do jack shit. Governments aren't gonna do jack shit. If this is the end of the world, fuck it. We might as well have a good time!

You're not wrong. And you're not an asshole.

Mike Primavera
@primawesome · Follow

My ability to dissociate has become too powerful. Now I'm just watching the fall of America like "hmm yeah that happens to empires" while I look for dog hats on Amazon.

6:45 PM · Jan 11, 2022

5.3K   Reply   Copy link

Read 61 replies

Nihilism, denialism, hedonism, and even apocalyptism are all reasonable responses when everything feels so hopeless. Especially in a country like the US where no one is talking about anything that actually matters, and our absurdist government is busy banning books and telling people where they can go to the bathroom. We can't save the Amazon and transition to a low energy future and get people to stop driving everywhere and build high-speed rails and fix capitalism and get people to start caring about each other again. We can't even ban assault weapons in America. Missouri just voted to defund public libraries. We're not even in the same stratosphere of the track we're supposed to be on. We couldn't even get people to wear a mask during a global fucking pandemic. I get it. Why bother doing something when there's nothing to be done? Might as well crack that six-a.m. beer and enjoy our front row seats to the end of the world.

But as hard as it is, and as easy as it is to give up or give in, we have to get comfortable with being in the shit. We have to know that we're probably not going to see much change in our lifetimes while remembering that even tiny bits of change still matter. Even if one thing goes off the rails, it doesn't mean it wasn't worth it.

There's a concept known to outdoorsman as "expedition behavior." How well do you hang in the shit? How do you handle adversity? When your kayak flips over and you lose all your gear do you give up and say everything is hopeless and wait for someone to rescue you? Or do you realize the situation is what it is and start figuring things out?

The thing about humans is we're antifragile. We actually get stronger when things get harder. We're all programmed not just to make it through all this (whatever this ends up looking like) but to thrive on the other side. And if we can remember that about ourselves, we can help more people recognize the same.

It's not all or nothing. You don't have to devote every waking moment to saving the world. You don't have to live with the weight of human extinction on your shoulders every day. You don't have to only pursue an idea if it's going to fix everything immediately. And you can still have a good time. You can have a great fucking time. I know I am!

I know that there are no easy solutions. I know everything I've talked about in this book is incredibly complex. Every solution I proposed is rife with externalities, so now I probably need to talk to an ecologist and an economist and a geopolitical strategist, and a biologist to understand all those potential downstream effects. Nothing is simple; everything is nuanced. And fighting our little pea-brain desire to put everything into black and white, right and wrong, blue and red, is one of the first steps to having a chance at fixing it.

But these changes are coming whether or not we do anything at all. The only thing that's certain in all this is that nothing is certain. None of us knows how any of this is going to play out.

We could fix all the problems of the world and still get hit by an asteroid tomorrow. We live on a tiny, fragile ball hurtling through the vast nothingness and terrible vacuum of space protected by an atmosphere that's no thicker than an apple peel relative to the size of our planet. The dinosaurs thrived for 165 million years before being completely wiped out by a stroke of bad luck.

But that's also the best part. It could go an infinite number of ways. We don't just need to accept uncertainty, we should be in wonder of it.

When we look to the future, as dystopian as it often seems, and as dystopian as it may actually be, there is always hope. As Jamie Wheal says: we're in the long disaster. We have to find durable, radical hope in something we can't see. We have to believe it's there. Because it is. But if we don't work together toward finding it, we never will.

Daniel Schmachtenberger once shared a story about talking with a new colleague of his who asked him, "When did you become post-tragic?"*

And he said, "What's your framework for that?" Because apparently that's how people talk.

Becoming post-tragic is the point after the nihilism and the hedonism and the apocalyptism. It's when you arrive at the other side. Somehow, some way, you realize that there isn't any other point—this is the only hill to die on. If I have to spend every ounce of energy hopefully saving a potentially unsavable thing ... then that still makes sense because it's literally the only thing to do. If we don't, there's nothing.

For all the doomsday prophecies I've laid out in this book, it's not about predicting the end of the world. It's about hope. It's about the fact that we are an incredible species. We have innovated a million things that no Malthus or Ehrlich could see when they were prophesizing the coming apocalypse. And I truly believe that it's still possible. We just have to get post-tragic. We have to start seeing the world in terms of difficult possibility instead of blindly believing in technology or throwing our hands in the air at the futility of our own destruction. We have to accept that it's going to take some fucking work. And it's going to take most of us to do it.

---

* Can you tell I'm a little bit in love with Daniel Schmachtenberger yet? Hi Daniel!

Hard Times Call for Furious Dancing

Every morning I awake torn between a desire
to save the world and an inclination to savor
it. This makes it hard to plan the day.

—E.B. White

In the face of this unthinkable, seemingly unsolvable tragedy—in the face of the end of the world as we know it—hedonism *is* a reasonable response. Shit is hard. And it's only going to get harder. But if we spend our whole lives working, whether it's to pay the bills or save the world, we just might forget to savor the world we're trying to save.

Not only *can* you have a good time, you fucking have to. As E.B. White continues: "If we forget to savor the world, what possible reason do we have for saving it? In a way, the savoring must come first."

The harder shit is, the more savoring we need to do. You have to find the joy that is everywhere on this incredible fucking planet. You have to find the thing that connects you not only to others, but to life itself. You have to break out of the habits of work and kids and dinner and TV and bed and do whatever it is that brings you joy. And the best part is, most of those most joyful things are free. The answer has always been and will always be right in front of us and all around us. We are incredibly social beings who need community and music and connection and release. Kalahari bushmen dance out their microtraumas, which leads to forgiveness between tribe members. Instead of holding onto grudges, or talking shit about fucking Linda, or withdrawing into the safety of ourselves, or spending our lives alone scrolling Instagram, shouldn't we all be dancing out our microtraumas as often as we can? As Jamie Wheal says in his book, *Recapture the Rapture*:

> Life is irreducibly tragic—we know that much. That's where healing becomes so essential—it gives us a chance to patch our bones and mend as we go onward. But occasionally, it's undeniably magic—and that's easier to forget. That's what inspiration does: It reminds us that there's beauty and perfection around us, if we only remember where to look.

We have to remember to look for that joy and beauty and healing and catharsis wherever we can find it. Whether it's your kids or your dogs or a beach or a mountain or just sitting under a blanket reading a book, you have to find the thing worth savoring that makes it all worth saving. Wheal recommends going to electronic music festivals to dance out your traumas and foster the connection in the same way that church has been doing for centuries. It turns out listening to music in large groups of people does something special to us. It connects us and elevates us and gives us access to a peak state that heals us. Something tells me both Taylor Swift and Beyoncé are providing this peak state at their concerts this year. Wheal also recommends "hedonic calendaring" with various substances like psychedelic mushrooms and MDMA and advanced breathing techniques and even some wild sex stuff. You don't have to do drugs or get into BDSM, but if it takes a mushroom trip to reconnect yourself with nature and the oneness of everything, I'm here for it. I've been doing psychedelics every few years since high school, and everything they say about their ability to help you transcend your ego, reconnect with the universe, and provide humility, healing, release, and connection is a hundred percent true. It's the only time I've ever really been able to see outside of myself. And I'm a better person for it.

But you do you, boo.

However you get there, get there. However you find it—find it.

We're all broken sometimes in some ways. We all waver. We all struggle with uncertainty and doubt. There is no wrong way to respond to grief, to tragedy, to trauma. But with or without substances, we can mend where we're broken through connection. It's the very connection we've been missing this whole time. The opposite of addiction isn't sobriety, it's community. We all have traumas in our lives. Traumas in our personal lives that make it harder to be honest with ourselves. Traumas that make it hard to forge real, resilient relationships with the people in our lives. Traumas that make it harder to spend valuable time and energy working for some vague future we may never see. Traumas from facing the undeniable potential for disaster that lies ahead. Traumas from how futile it all seems when you start thinking about it. But you can exist with these. You can be sad and devastated and hopeful and energized and everything everywhere all at once. You can be sad and a little bit heartbroken every day, and you can still be happy right fucking now.

Permaculture expert and professor Andrew Millison, when asked on Nate Hagen's show what his one wish for the world would be if he had a magic lamp, responded: "I wish every human could have some sort of 'oneness with the universe' experience . . . where they shed their dramas and their worries and illusions and recognize their connection; that we're just all part of this fabric. To see their lives and themselves and creation for what they are."

Because once you see that connection and feel it, you can't unsee it. The underlying cause of all our problems is not perceiving from that wholeness first. The spontaneous right action we need to save the world emerges when you realize we're all one thing. You won't spend every moment of every day working to save the world; no one does. But you'll never forget that beauty and power and inspiration. You will always be tethered to that understanding. You can't forget it and you can't unlearn it. And while some people experience and recognize that beauty, it can also be easy to let it take you down the wrong path. It's so easy to believe that even though you see the truth of who we are and what everything is, the fact that so few other people do makes it hopeless. But I promise you, there is so much hope to be had.

I want to close with one last thought from Schmachtenberger.

Before a baby chick is hatched, as it lives inside of its egg, it survives in a sac of amniotic fluid. And every day while the little chick fetus is in there, that fluid gets lower and lower. And maybe that little chick is like, "Holy shit, what am I gonna do? I've been living on this juice, and it's almost gone, and I don't have anywhere to get anymore juice!"

But then one day, in the incredibly, endlessly efficient way that evolution and Mother Nature work, at the exact moment that the juice runs out, the chick is strong enough to peck through the shell and break on through to the other side. So maybe, right now, we're just stuck in the amniotic sac. Maybe we're so young and stupid we can't even see the constraints of the system we're about to break free from. But we have to keep working to break free. We have to start pecking at the shell if we're ever gonna get out.

It's not going to be easy. And more often than not, it's going to be real fucking hard. But in those inevitable times when you feel trapped and doomed, when you see the juice running out and it feels like we're living in an unsavable world, when you see all the people who don't see it and don't care and aren't changing a goddamn thing, and all your time and energy is being wasted on what feels like an utterly hopeless cause—it's time to start savoring so you can remember what we're trying to save. These hard times call for some furious fucking dancing.

A Profound Thanks

It's no secret that this book has been written on the shoulders of many people far smarter than I am. It's pretty much a greatest hits mixtape of the best things I've heard that changed my mind or inspired me, and I guess I hoped that putting them all together might inspire someone else too. It's not full of original ideas (though there are a few). The way I see my part in all this is making this information more accessible, more relatable, and bringing it to as many people as humanly possible.

Rather than write a sappy love letter to all of the people who made this book possible, I'm going to list the resources that were indispensable to me in creating it. I also highly recommend you read as many of these books as you can.

I wish I could thank my fact checker for saving me from some embarrassing mistake, but I'm too poor to afford a factchecker for a book with ten thousand facts in it, so apologies in advance if I got something wrong or completely fucked something up. I'm 100 percent sure I did, so please let me know if you catch something, whether glaring or minuscule.

It goes without saying, but I'll say it again: a resounding thank you and unpayable debt of gratitude to Nate Hagens and his podcast, *The Great Simplification*. If you haven't started listening, do. Much of this book was inspired by various guests on his show, including Art Berman, William Rees, Peter Whybrow, Jamie Wheal, Paul Ehrlich, Peter Ward, Herman Daly, Andrew Millison, and of course, Daniel Schmachtenberger. I highly recommend you give all of those episodes a listen.

You've already read plenty of insights and direct passages from the following books, but I recommend each and every one of them (though they vary in terms of readability and accessibility):

Peter Zeihan
The End of the World Is Just the Beginning: Mapping the Collapse of Globalization (Harper Business, 2022)

Robert Reich
Saving Capitalism: For the Many, Not the Few (Vintage, 2015)

Matthias Schmelzer, Aaron Vansintjan, Andrea Vetter
The Future is Degrowth: A Guide to a World Beyond Capitalism (Verso, 2022)

Charles Massy
The Call of the Reed Warbler: A New Agriculture, A New Earth (Chelsea Green Publishing, 2018)

Jerry Z. Muller
Adam Smith in His Time and Ours: Designing the Decent Society (Princeton University Press, 1995)

Jamie Wheal
Recapture the Rapture: Rethinking God, Sex, and Death in a World That's Lost Its Mind (Harper Wave, 2021)

Peter Whybrow
American Mania: When More Is Not Enough (W. W. Norton & Company, 2006)

Herman Daly
Valuing the Earth: Economics, Ecology, Ethics (MIT Press, 2nd edition, 1995)

And the only fiction addition to the list:
Kim Stanley Robinson
The Ministry for the Future
(Hachette B & Blackstone Publishing, 2020)

There are also plenty of books I've read over the years that helped to build my understanding of who we are as humans, helped develop and inform how I see the world, and helped me to sound less stupid than I was before I read them. Here are a few I recommend:

David Graeber and David Wengrow
The Dawn of Everything: A New History of Humanity (Farrar, Straus and Giroux, 2021)

James Suzman
Affluence without Abundance: The Disappearing World of the Bushmen (Bloomsbury USA, 2017)

Bill Gates
How to Avoid a Climate Disaster: The Solutions We Have and the Breakthroughs We Need (Vintage, 2021)

Richard Rothstein
The Color of Law: A Forgotten History of How Our Government Segregated America (Liveright, 2017)

Nicholas D. Kristof and Sheryl WuDunn
Tightrope: Americans Reaching for Hope (Vintage, 2020)

Hans Rosling
Factfulness: Ten Reasons We're Wrong About the World—and Why Things Are Better Than You Think (Flatiron, 2018)

Steven Pinker
Enlightenment Now: The Case for Reason, Science, Humanism, and Progress (Penguin, 2018)

And finally: THE INTERNET.

This book would have entailed years and years of painstaking research on my part without the miracle of the internet. It is the culmination of years and years of other people's painstaking research. Obviously, the division of labor and specialization and the information superhighway have made us faster and more efficient. Like I said, I'm not a Luddite!

Notable mention goes to: Our World in Data—seriously what an incredible, free resource that gives me hope for the future of humanity. Almost any random chart I googled was already available, beautifully formatted, with the underlying data sets available for download and an informative article/analysis to boot. Hannah Ritchie especially seems to be interested in all the same topics as I am. Thank you, Hannah.

And thank you to the US Energy Information Administration, The International Energy Agency, the UN, the US Bureau of Labor Statistics, The World Bank, and basically every newspaper and magazine and journalist doing the work to make information accessible.

★★★

Lastly, if you're reading this . . . thank you. I put hundreds upon hundreds of hours into writing, researching, designing, editing, and publishing this book in the hopes that I could reach anyone at all. Many friends, and even family members, gave up halfway through. I've been told it's too dark, too hard, too bleak, and even too silly. But if you've come this far, maybe you're willing to do me one last solid and leave a review on Amazon. Those reviews will help get this book in the hands of people who need to read it. It would mean the world to me.

Knowledge is power, guys. Now let's go save the world.

Resources

Energy Conversions

Joule: The energy required to lift one newton (102 g or a small apple) one meter against Earth's gravity.

Watt: One joule per second. A human climbing a flight of stairs is doing work at the rate of about 200 watts.

BTU (British Thermal Unit): The amount of heat required to raise the temperature of one pound of water by one degree Fahrenheit.

1 joule = 0.0009478 BTU
1 BTU = 1,055.06 joules = 0.252 calories
1 kilojoule = 1,000 joules = 0.947817 BTU
1 calorie = .00397 BTU
1 kilocalorie = 3.968 BTU
1 BTU = 1.055 kilojoules
1 kilojoule = 0.239 kilocalories
1 kilocalorie = 4.1868 kilojoules
1 watthour = 3,600 joules
1 kilowatt-hour = 3.6 megajoules = 3,412 BTU
1 megajoule = 1 million joules

1 horsepower = 2,545 BTU per hour
1 watt = 1 joule per second
1 watt = 3.412 BTU per hour
1 BTU per hour = 0.293 watts
1 kilowatt = 1,000 watts
1 kilowatt = 3,412 BTU per hour
1 kilowatt = 1.341 horsepower (electric)
1 horsepower (electric) = 0.7457 kilowatts
1 megawatt = 1 million watts
1 gigawatt = 1 billion watts
1 terawatt = 1 trillion watts

Average Energy Content of Various Fuels

1 cubic foot of natural gas = 1,008–1,034 BTU

1 gallon of crude oil = 138,095 BTU

1 barrel of crude oil = 5,800,000 BTU

1 gallon of residual fuel oil = 149,690 BTU

1 gallon of gasoline = 125,000 BTU

I gallon of kerosene = 135,000 BTU

1 gallon of middle distillate or diesel fuel oil = 138,690

1 gallon of liquefied petroleum gas (LPG) = 95,475 BTU

1 pound of coal = 8,100–13,000 BTU

1 ton of coal = 16,200,000–26,000,000 BTU

1 ton of wood = 9,000,000–17,000,000 BTU

Average U.S. Household Electric Consumption for Major Electrical Appliances

Appliance	Annual kWh Consumption
Central Air Conditioning	2,667
Room Air Conditioning	738
Water Heater	2,671
Refrigerator	1,155
Freezer	1,204
Range / Oven	458
Dishwasher	299
Clothes Washer	99
Clothes Dryer	875

End of the World as We Know It
BINGO

1.5° warming reached	We ask AI to save the planet and it kills us all because we are the parasites	2nd civil war in America	Global financial collapse after first major economy defaults on debt	Yellowstone erupts, triggering nuclear winter
AMOC breaks down	National electric grid fails	Thawing of Siberian permafrost releases some crazy ancient disease and we have another pandemic	Hackers take over the American nuclear arsenal	Bee extinc-tion
Canadian lakes turn to jelly	Warming planet leads to fungus taking over the human brain turning us into flesh eating monsters	FREE	$20 a gallon gas	Solar flare knocks out global communication networks
Bezos flies to Mars	Global sea levels rise two feet	Russia invades Europe	US stops protecting global shipping corridors causing the return of piracy and the collapse of globalization	Uber wealthy people retreat to their bunkers
Cannibalistic giants cause a trophic cascade leading to oceanic ecosystem breakdown	Orcas chew and destroy communica-tion cables under the sea (likely out of spite)	Americans get universal healthcare	First country to fully decivilize and return to an agrarian economy	Actual nuclear war

Notes

The Dumpster Fire We Started

The Walking Worried

Epigraph: Charles M. Schulz, adapted from Peanuts, November 26, 1964. https://www.facebook.com/schulzmuseum/photos/a.110604408053/1015782 2839458054/?type=3.

The Incredible Everything Machine

Epigraph: Jamie Wheal, Recapture the Rapture: Rethinking God, Sex and Death in a World That's Lost its Mind (Harper Wave, 2021).

Image 1.1: 1900s photograph of burlaks on the Volga River. From Rybinsk state historical-architectural art museum and national park photofiles. Public domain.

✦✦✦

1 Michael Ruhlman, *Grocery: The Buying and Selling of Food in America* (Abrams Press, 2017).

2 I can't find the original study, but Nordhaus's findings are summarized in this BBC article: "Why the falling cost of light matters," BBC News, February 6, 2017, https://www.bbc.com/news/business-38650976.

3 Paul R. Liegey, "Microwave Oven Regression Model," U.S. Bureau of Labor Statistics, Consumer Price Index, last modified October 16, 2001, https://www.bls.gov/cpi/quality-adjustment/microwave-ovens.htm.

4 W. Michael Cox, Richard Alm, "Time Well Spent: The Declining Real Cost of Living in America," Federal Reserve Bank of Dallas 1997 Annual Report, Accessed April 11, 2023, https://www.dallasfed.org/~/media/documents/fed/annual/1999/ar97.pdf

5   Thomas R. Tibbets, "Expanding Ownership of Household Equipment," *Monthly Labor Review* 87, no. 10 (1964): 1131–37, http://www.jstor.org/stable/41835526.

6   James D. Lutz, "Lest We Forget, a Short History of Housing in the United States," Berkeley Lab, Energy Technologies Area, August 2004, https://www.aceee.org/files/proceedings/2004/data/papers/SS04_Panel1_Paper17.pdf.

The Story of Everything

Epigraph: Unknown.

Figure 1.2: "Daily per capita energy consumption across social structures," Taylor Ahlstrom. Data source: "Human energy use (endosomatic/exosomatic)," Environmental Justice Organization, Liabilities and Trade, accessed March 31, 2023.

★★★

7   Marvin Harris, *Cultural Materialism: The Struggle for a Science of Culture* (Vintage, 1980).

8   Hiram M. Drache, "The Impact of John Deere's Plow," Northern Illinois University, 2001, https://www.lib.niu.edu/2001/iht810102.html.

9   "Human energy use (endosomatic/exosomatic)," Environmental Justice Organization, Liabilities and Trade, accessed March 31, 2023, http://www.ejolt.org/2012/12/human-energy-use-endosomatic-exosomatic.

How many licks to the Tootsie Roll center of a Tootsie pop?

Epigraph: Dipesh Chakrabarty, *The Climate of History in a Planetary Age* (University of Chicago Press, 2021).

★★★

10   "What is a Human Being Worth (in Terms of Energy)?" The Oil Drum: Europe, July 20, 2008, http://theoildrum.com/node/4315.

11   "How much electricity does an American home use?" U.S. Energy Information Administration, last modified October 12, 2022, https://www.eia.gov/tools/faqs/faq.php?id=97&t=3.

12   "Fossil fuel sources accounted for 79% of U.S. consumption of primary energy in 2021," U.S. Energy Information Administration, July 1, 2022, https://www.eia.gov/todayinenergy/detail.php?id=52959.

The Boy Who Cried Peak Oil

Epigraph: Jeff Rubin, *Why Your World Is About to Get a Whole Lot Smaller: Oil and the End of Globalization* (Random House, 2009).

Figure 2.1: "US Oil production vs. Hubbert upper-bound curve," adapted from "Hubbert Upper-Bound Peak 1956" by Plazak, licensed under CC Attribution 4.0

Image 2.2: *Beverly Hillbillies*, public domain.

Image 2.3: "Fracking: Hydraulic Fracturing," illustration by Vectormine, commercially licensed.

★★★

13   Jack Rivers, Stuart Lewis, "2022 exploration drilling review and 2023 High Impact well (HIW) drilling outlook," S&P Global Commodity Insights, December 15, 2022, https://www.spglobal.com/commodityinsights/en/ci/research-analysis/2022-exploration-drilling-review-and-2023-high-impact-well-hiw.html.

14   "Monthly Crude Oil and Natural Gas Production," U.S. Energy Information Administration, May 31, 2023, https://www.eia.gov/petroleum/production/#oil-tab.

15   "Production Decline Curve Analysis in the Annual Energy Outlook 2023," U.S. Energy Information Administration, March 16, 2023, https://www.eia.gov/analysis/drilling/curve_analysis/.

16   Megan C. Guilford et al., "A New Long Term Assessment of Energy Return on Investment (EROI) for U.S. Oil and Gas Discovery and Production," *Sustainability* 3, no. 10, (October 2011): 1866-1887, https://doi.org/10.3390/su3101866.

17   David J. Murphy, "The implications of the declining energy return on investment of oil production," *Philosophical Transactions of the Royal Society A* 372 (2006), http://doi.org/10.1098/rsta.2013.0126.

18   C.W. King, "Comparing World Economic and Net Energy Metrics, Part 3: Macroeconomic Historical and Future Perspectives," *Energies* 8, no. 11 (2015): 12997–13020, https://doi.org/10.3390/en81112348.

Hold onto Your Butts

Epigraph: *Jurassic Park*, directed by Steven Spielberg, featuring Samuel L. Jackson, Universal Pictures, 1993.

Figure 3.1: "Global GDP vs. Energy Consumption, 1965–2015," Taylor Ahlstrom, Data source: BP Statistical Review of World Energy, 200; Our World in Data.

Figure 3.2: "Share of energy consumption by source, United States," Hannah Ritchie, Max Roser, and Pablo Rosado (2022) – "Energy," OurWorldInData. org, https://ourworldindata.org/energy.

Figure 3.3: "Supply Chain of Raw Materials Used in the Manufacturing of Light-Duty Vehicle Lithium-Ion Batteries," Igogo, Tsisilile, Debra Sandor, Ahmad Mayyas, and Jill Engel-Cox, 2019, Golden, CO: National Renewable Energy Laboratory, NREL/TP-6A20-73374. https://www.nrel.gov/docs/fy19osti/73374.pdf.

Figure 3.4: "World population with and without synthetic nitrogen fertilizers," Hannah Ritchie, OurWorldInData.org, https://ourworldindata.org/how-many-people-does-synthetic-fertilizer-feed.

★★★

19   "A number of countries have decoupled economic growth from energy use, even if we take offshored production into account," Our World in Data, November 30, 2021, https://ourworldindata.org/energy-gdp-decoupling.

20   "Are electric vehicles definitely better for the climate than gas-powered cars?" MIT Climate Portal, October 13, 2022, https://climate.mit.edu/ask-mit/are-electric-vehicles-definitely-better-climate-gas-powered-cars.

21   Cameron Tarry, Faith Martiniez-Smith, "Supply Chain for Lithium and Critical Minerals Is … Critical," Clearpath, June 11, 2020, https://clearpath.org/tech-101/supply-chain-for-lithium-and-critical-minerals-is-critical/.

22   Peter Zeihan, *The End of the World Is Just the Beginning: Mapping the Collapse of Globalization* (Harper Business, 2022).

23   Mark P. Mills, "Mines, Minerals, and 'Green' Energy: A Reality Check," Manhattan Institute, July 9, 2020, https://www.manhattan-institute.org/mines-minerals-and-green-energy-reality-check.

24   "Rare Earth Magnet Market Outlook to 2035," Adamas Intelligence, April 20, 2022, https://www.adamasintel.com/rare-earth-magnet-market-outlook-to-2035/.

25   Hannah Ritchie, "How does the land use of different electricity sources compare?" Our World in Data, June 16, 2022, https://ourworldindata.org/land-use-per-energy-source.

26   Blaine Friedlander, "Touted as clean, 'blue' hydrogen may be worse than gas or coal," Cornell Chronicle, August 12, 2021, https://news.cornell.edu/stories/2021/08/touted-clean-blue-hydrogen-may-be-worse-gas-or-coal.

27   Nicola Warwick, Paul Griffiths, James Keeble, et al., "Atmospheric implications of increased hydrogen use," University of Cambridge and National Center for Atmospheric Sciences, University of Reading, April 8, 2022, https://www.gov.uk/government/publications/atmospheric-implications-of-increased-hydrogen-use

28   Mickey Francis, "About 7% of fossil fuels are consumed for non-combustion use in the United States," *Today in Energy*, U.S. Energy Information Administration, April 6, 2018, https://www.eia.gov/todayinenergy/detail.php?id=35672.

29   Hannah Ritchie, "How many people does synthetic fertilizer feed?" Our World in Data, November 7, 2017, https://ourworldindata.org/how-many-people-does-synthetic-fertilizer-feed.

30   Leigh Krietsch Boerner, "Industrial ammonia production emits more CO_2 than any other chemical-making reaction. Chemists want to change that," *Chemical & Engineering News*, June 15, 2019, https://cen.acs.org/environment/green-chemistry/Industrial-ammonia-production-emits-CO2/97/i24.

The Perpetual Motion Machine

Epigraph: Milton Friedman, *Donahue*, Phil Donahue, episode aired in 1979 on WGN.

Figure 4.1: "Global gross domestic product (GDP) at current prices from 1985 to 2028," Taylor Ahlstrom. Data source: IMF World Economic Outlook Database.

Figure 4.2: "Global GDP over the last two millennia," published online at OurWorldInData.org, https://ourworldindata.org/grapher/world-gdp-over-the-last-two-millennia.

Image 4.3: Anat Shenker-Osorio (@anatosaurus), Twitter, March 24, 2020, https://twitter.com/anatosaurus/status/1242237957332856838.

Image 4.4: "Get in loser, we're going shopping," original meme by Taylor Ahlstrom. Source material: *Mean Girls*, directed by Mark Waters, featuring Rachel McAdams, Lacey Chabert, Amanda Seyfried, Paramount Pictures, 2004. John Maynard Keynes, public domain.

✦✦✦

31   Texas Lieutenant Governor Dan Patrick, interview by Tucker Carlson, *Tucker Carlson Tonight*, March 23, 2020.

32   Nicholas Georgescu-Roegen, *Energy and Economic Myths: Institutional and Analytical Economic Essays* (Pergamon Press, 1976).

Infinite Growth in a Finite System

Epigraph: Nate Hagens, "Part II: The Human Superorganism," *The Great Simplification*, https://youtu.be/oNewKEOby80.

Figure 5.1: "Circular flow of goods and income," by Irconomics, licensed under CC Attribution Share Alike 3.0, fonts updated, https://en.wikipedia.org/wiki/Circular_flow_of_income#/media/File:Circular_flow_of_goods_income.png

Image 5.2: "The Whole Fucking Planet," Taylor Ahlstrom, created with original artwork from Macrovector on Freepik, commercially licensed, https://www.freepik.com/author/macrovector.

✦✦✦

33   Joss Tantram, "Entropic Overhead: Measuring the circular economy," terrafiniti, September 2013, https://www.terrafiniti.com/entropic-overhead-measuring-the-circular-economy/.

34   Marian L. Tupy, Gale L. Pooley, *Superabundance: The Story of Population Growth, Innovation, and Human Flourishing on an Infinitely Bountiful Planet* (Cato Institute, 2022).

35   Mario Pagliaro, Franceso Meneguzzo, "Lithium battery reusing and recycling: A circular economy insight," *Heliyon* 5, no. 6 (2019), https://doi.org/10.1016/j.heliyon.2019.e01866.

Overshoot

Epigraph: Herman Daly, "Developing Economies and the Steady State," *Valuing the Earth, 2ⁿᵈ Edition, Economics, Ecology, Ethics* (MIT Press, 1993).

Figure 6.1: "Changing distribution of the world's land mammals," Hannah Ritchie. *Published online at OurWorldInData.org*, https://ourworldindata.org/wild-mammals-birds-biomass.

Figure 6.2: "Cereal yield vs. fertilizer use, 2020," *published online at OurWorldInData.org*, https://ourworldindata.org/grapher/cereal-crop-yield-vs-fertilizer-application.

Figure 6.3: "Annual waste in millions of metric tons (2021)," Taylor Ahlstrom. Source: World Bank, "Trends in Solid Waste Management;" Global Carbon Project (2023).

★★★

36   "Global Issues: Population," United Nations, Accessed April 13, 2023, https://www.un.org/en/global-issues/population

37   Mathis Wackernagel, Niels B. Schulz, Diana Deumling, et al., "Tracking the ecological overshoot of the human economy," *PNAS* 99, no. 14 (June 2002), https://www.pnas.org/doi/10.1073/pnas.142033699.

38   Hannah Ritchie, Klara Auerbach, "Wild mammals make up only a few percent of the world's mammals," Our World in Data, December 15, 2022, https://ourworldindata.org/wild-mammals-birds-biomass.

39   Ron Milo, interview by Damian Carrington, "Humans just 0.01% of all life but have destroyed 83% of wild mammals," *The Guardian*, May 21, 2018, https://www.theguardian.com/environment/2018/may/21/human-race-just-001-of-all-life-but-has-destroyed-over-80-of-wild-mammals-study.

40   "Crop yields, World, 1961 to 2021," Our World in Data, Accessed April 13, 2021, https://ourworldindata.org/grapher/key-crop-yields

41   Patricio Grassini, Kent M. Eskridge, Kenneth G. Cassman, "Distinguishing between yield advances and yield plateaus in historical crop production trends," *Nature Communications* 4, 2918 (December 2013), https://doi.org/10.1038/ncomms3918.

42   Gittemarie, "The Environmental Impact of Lab Grown Meat – and why we need it," Updated December 9, 2022, Gittemary, https://www.gittemary.com/2022/12/the-environmental-impact-of-lab-grown-meat-and-why-we-need-it.html

43   "What a Waste 2.0: Trends in Solid Waste Management," The World Bank, accessed March 31, 2023, https://datatopics.worldbank.org/what-a-waste/trends_in_solid_waste_management.html.

44   Hannah Ritchie, "How much of global greenhouse gas emissions come from food?" Our World in Data, March 18, 2021, https://ourworldindata.org/greenhouse-gas-emissions-food

How bad could it be, really?

Pretty Fucking Bad

Epigraph: Greta Thunberg, "Our House is on Fire," speech to World Economic Forum, Davos, 2019.

Figure 7.1: "The Keeling Curve," Adapted from Scripps Institution of Oceanography at UC San Diego, licensed under CC BY license, Creative Commons Attribution 4.0 International License, https://keelingcurve.ucsd.edu/.

★★★

45   Andrew Moseman, "Why do we compare methane to carbon dioxide over a 100-year timeframe? Are we underrating the importance of methane emissions?" MIT Climate Portal, June 28, 2021, https://climate.mit.edu/ask-mit/why-do-we-compare-methane-carbon-dioxide-over-100-year-timeframe-are-we-underrating

46   Bill McKibben, "Global Warming's Terrifying New Math," *Rolling Stone*, July 19, 2012, https://www.rollingstone.com/politics/politics-news/global-warmings-terrifying-new-math-188550/.

47   Robert B. Jackson, "The carbon budget for 1.5°C," Global Carbon Project, December 15, 2019, https://www.globalcarbonproject.org/global/pdf/carbonbudget/1.5C_Animation_Jackson_GCP_Stanford_2019.pdf

48   "State of the Global Climate 2022," World Meteorological Organization, 2023, https://library.wmo.int/index.php?lvl=notice_display&id=22265#.ZEKI3-zMKcL.

The Short Term

Epigraph: Sir Richard Mottram, overheard speaking to a colleague in British House of Commons, March 2002.

Figure 7.2: "Crude Oil vs Gasoline Prices," Macrotrends, https://www.macrotrends.net/2501/crude-oil-vs-gasoline-prices-chart.

Image 7.3: "Pivot!" original meme by Taylor Ahlstrom. Source: *Friends*, episode 113, "The One with the Cop," directed by Andrew Tsao, featuring David Schwimmer, aired February 25, 1999 on NBC. Photo: Evergreen

Image 7.4: "Pears grown in Argentina, packed in Thailand," meme adapted by Taylor Ahlstrom. Source material: u/Donatello, "[Request] Is it really more economically viable to ship Pears Grown in Argentina to Thailand for packing?" https://www.reddit.com/r/theydidthemath/comments/1131cx3/request_is_it_really_more_economically_viable_to/. Source 2: https://www.marinetraffic.com/en/voyage-planner.

Image 7.5: "You're Not Wrong, Walter," meme adapted by Taylor Ahlstrom. Source material: *The Big Lebowski*, directed by Joel and Ethan Cohen, featuring Jeff Bridges, Working Title Films, 1998.

Image 7.6: "Skeptical third-world child," meme adapted by Taylor Ahlstrom. Source: u/manute3392, photo credit: u/Nepalm, June 21, 2012, https://www.reddit.com/r/pics/comments/vds3u/skeptical_3rdworld_child/?rdt=56704Reddit.com.

★★★

49   "Agricultural Trade," Economic Research Service, US. Department of Agriculture, last modified February 22, 2023, https://www.ers.usda.gov/data-products/ag-and-food-statistics-charting-the-essentials/agricultural-trade/.

50   "Food Outlook – Biannual Report on Global Food Markets," Food and Agriculture Organization of the United Nations, November 2022, Rome, https://doi.org/10.4060/cc2864en.

51   Daniel Much, "New AFBF Survey Shows Drought's Increasing Toll on Farmers and Ranchers," American Farm Bureau Federation, August 14, 2022, https://www.fb.org/market-intel/new-afbf-survey-shows-droughts-increasing-toll-on-farmers-and-ranchers.

52   Alex Findijs, "Federal officials warn of impending water crisis in the American Southwest," World Socialist Web Site, June 23, 2022, https://www.wsws.org/en/articles/2022/06/24/fkkg-j24.html.

53   Margaret Osborne, "A Century Ago, This Water Agreement Changed the West. Now, the Region Is in Crisis," Smithsonian Magazine, November 28, 2022, https://www.smithsonianmag.com/smart-news/a-century-ago-this-water-agreement-changed-the-west-now-the-region-is-in-crisis-180981169/.

54   "Turning the Tide: A Call to Collective Action," Global Commission on the Economics of Water, March 2023, https://turningthetide.watercommission.org/.

55   Viviane Clement, Kanta Rigaud, et al., "Groundswell Part 2: Acting on Internal Climate Migration," World Bank, Washington, DC, 2021, https://openknowledge.worldbank.org/entities/publication/2c9150df-52c3-58ed-9075-d78ea56c3267.

56   "Immigration," Gallup, accessed March 31, 2023, https://news.gallup.com/poll/1660/immigration.aspx.

57   "Southeast Asia Energy Outlook 2022: Key Findings," International Energy Agency, accessed March 31, 2023, https://www.iea.org/reports/southeast-asia-energy-outlook-2022/key-findings.

58   Alan Buis, "Too Hot to Handle: How Climate Change May Make Some Places Too Hot to Live," NASA Climate, March 9, 2022, https://climate.nasa.gov/ask-nasa-climate/3151/too-hot-to-handle-how-climate-change-may-make-some-places-too-hot-to-live/.

59   "Emerging 'Extreme Heat Belt' will Impact Over 107 Million Americans by 2053," First Street Foundation, August 15, 2022, https://firststreet.org/press/press-release-2022-heat-model-launch/.

60   Masashi Takahashi, "Heat stress on reproductive function and fertility in mammals," *Reproductive Medicine and Biology* 11, no. 1, (January 2012): 37-47, https://doi.org/10.1007/s12522-011-0105-6.

61   Jennifer Holleis, "How climate change paved the way to war in Syria," Deutsche Welle, February 26, 2021, https://www.dw.com/en/how-climate-change-paved-the-way-to-war-in-syria/a-56711650.

62   Paul Iddon, "As the death toll falls and talks progress, could Syria finally know peace in 2023?" Arab News, January 14, 2023, https://www.arabnews.com/node/2232556/middle-east.

63   Jan Selby, Omar S. Dahi, Christiane Frölich, Mike Hulme, "Climate change and the Syrian civil war revisited," *Political Geography* 60, (September 2017): 232–234, https://doi.org/10.1016/j.polgeo.2017.05.007.

The Long Term

Epigraph: William Catton, Jr., *Overshoot: The Ecological Basis of Revolutionary Change* (University of Illinois Press, 1982).

Figure 8.1: "Ocean surface currents," Dr. Michael Pidwirny, August 10, 2007, U.S. Government, public domain, https://commons.wikimedia.org/wiki/File:Corrientes-oceanicas.png

Figure 8.2: "Franklin Gulf Stream," Benjamin Franklin and James Poupard, 1786, Library of Congress, public domain.

Image 8.3: "So I got that going for me," meme adapted by Taylor Ahlstrom. Source material: *Caddyshack,* directed by Harold Ramis, featuring Bill Murray, Warner Bros. Pictures, 1981.

Image 8.4: Tracey Thorn (@tracey_thorn), Twitter, May 23, 2019, https://twitter.com/tracey_thorn/status/1131576299179388928.

Image 8.5: "Who had gelatinous lakes?" meme adapted by Taylor Ahlstrom. Source material: *Cabin in the Woods,* directed by Drew Goddard, featuring Richard Jenkins, Lions Gate, 2012.

Image 8.6: "Who had cannibalistic giants?" meme adapted by Taylor Ahlstrom. Source material: *Cabin in the Woods,* directed by Drew Goddard, featuring Richard Jenkins, Lions Gate, 2012.

★★★

64   Paul Webb, "9.2 The Gulf Stream," *Introduction to Oceanography*, Roger Williams University, accessed March 31, 2023, https://rwu.pressbooks.pub/webboceanography/chapter/9-2-the-gulf-stream/.

65   Kristyn Ecochard, "What's causing the poles to warm faster than the rest of Earth?" NASA, April 6, 2011, https://www.nasa.gov/topics/earth/features/warmingpoles.html.

66   L. Caesar, G.D. McCarthy, D.J.R. Thornalley, et al., "Current Atlantic Meridional Overturning Circulation weakest in last millennium," *Nature Geoscience* 14, (2021):118–120, https://doi.org/10.1038/s41561-021-00699-z.

67   Stuart Scott Warren, Austin Stuart Hart, Isaac Saarman, et al., "Hydrogen Sulfide And Arsenic-rich Waters Originating From Buried Wood Waste Along The Commencement Bay Waterfront, Tacoma, Washington," *Geological Society Of America* 39, no. 44, (2007): 76–76, https://soundideas.pugetsound.edu/faculty_pubs/2107/

68   "Issues Brief: Ocean deoxygenation," The International Union for Conservation of Nature, accessed March 31, 2023,   https://www.iucn.org/resources/issues-brief/ocean-deoxygenation.

69   Adele M. Dixon, Piers M. Forster, et al., "Future loss of local-scale thermal refugia in coral reef ecosystems," *PLOS Climate* 1, no. 2 (2022), https://doi.org/10.1371/journal.pclm.0000004.

How Did We Get Here?

Sounds Like an Anthropocene Problem

Epigraph: "Anti-Hero," track no. 3 on *Midnights*, Republic, 2022, Taylor Swift.

Image 9.1: "Thank you Jesus for this food," meme adapted by Taylor Ahlstrom. Source: iStock images; "Family Praying," commercially licensed. Source Image 2: Jack Kurtz, "Onion Harvest," May 20, 1999, commercially licensed.

Image 9.2: Quora, July 12, 2023, https://www.quora.com/Why-do-some-people-cruelly-kill-animals-for-meat-instead-of-just-going-to-the-supermarket-to-get-it.

Image 9.3: Sophia (@pastoralcomical), Twitter, Nov 11, 2021, https://twitter.com/pastoralcomical/status/1458916126256160778.

★★★

70   Charles Massy, *Call of the Reed Warbler: A New Agriculture A New Earth*, (University of Queensland Press, 2017).

Sounds Like the Oxen's problem

Epigraph: The effing bible

Image 9.4: "Medieval illustration of men harvesting wheat with reaping-hooks or sickles," c. 1310, public domain, https://en.wikipedia.org/wiki/Serfdom#/media/File:Reeve_and_Serfs.jpg.

★★★

71   Guido Alfani, "Economic Inequality in Preindustrial Times: Europe and Beyond." *Journal of Economic Literature*, 59, no. 1 (March 2021): 3–44. https://doi.org/10.1257/jel.20191449.

72   Larry Elliott, "World's eight richest people have same wealth as poorest 50%," *The Guardian*, Jan 16, 2017, https://www.theguardian.com/global-development/2017/jan/16/worlds-eight-richest-people-have-same-wealth-as-poorest-50.

73   "Richest 1% bag nearly twice as much wealth as the rest of the world put together over the past two years," Oxfam, January 16, 2023, https://www.oxfam.org/en/press-releases/richest-1-bag-nearly-twice-much-wealth-rest-world-put-together-over-past-two-years.

74   "Why Medieval Serfs Had More Vacation Time Than You Do Today," Curious Historian, October 20, 2019, https://curioushistorian.com/why-medieval-serfs-had-more-vacation-time-than-you-do-today

Sounds Like a Personal Problem

Epigraph: Unknown.

★★★

75   Edward G. Ryan, chief justice of Wisconsin Supreme Court, address to graduating class at University of Wisconsin, 1873.

Sounds Like a Tomorrow Problem

Epigraph: Martin Luther King, Jr., "Riverside Church Speech," April 4, 1967.

★★★

76   Arianne J. van der Wal, Hannah M. Schade, Lydia Krabbendam, et al., "Do natural landscapes reduce future discounting in humans?" *Proceedings of the Royal Society B* 280, no. 1773, (December 2013), http://doi.org/10.1098/rspb.2013.2295.

Working with What We've Got

Heart, Will, and Mind

Epigraph: Immanuel Kant, "Idea for a Universal History with a Cosmopolitan Purpose," 1784.

Image credit: Isiah (@forevertawl), Twitter, August 15, 2020, https://twitter.com/forevertawl/status/1294424061590867968.

★★★

77   Robert Reich, *Saving Capitalism: or the Many, Not the Few* (Vintage, 2016).

78   Scott Mautz, "A Biologist's Study of 'Superchickens' Reveals a Surprising Truth for Leaders," Inc., February 2017, https://www.inc.com/scott-mautz/a-biologists-study-of-superchickens-reveals-a-surprising-truth-for-leaders.html.

79   Jordan Moss, Peter J. O'Connor, "The Dark Triad traits predict authoritarian political correctness and alt-right attitudes," *Heliyon* 6, no. 7, (July 2020), https://doi.org/10.1016%2Fj.heliyon.2020.e04453

The Benevolent Dictator

Epigraph: Senator John Sherman, congressional speech, 1890.

Figure 10.1: "Map of North Carolina's 12th Congress," October 30, 2004, public domain, https://commons.wikimedia.org/w/index.php?curid=2721125.

Image 10.2: Edward Stockwell (@EdwardStockwell), Twitter, August 4, 2021, https://twitter.com/EdwardStockwell/status/1422716134479044608.

Image 10.3: "I am time itself." I was unable to find the original poster of this meme. So, I'm sorry, but thank you for your fridge photo.

Image 10.4: "It's one banana, Michael," meme adapted by Taylor Ahlstrom. Source credit: *Arrested Development*, episode 113, "Charity Drive," directed by Greg Mottola, featuring Jessica Walter, aired November 30, 2003 on Fox.

Image 10.5: "Crying Indian," Keep America Beautiful, Inc., 1971.

Figure 10.6: "Average federal income tax spending, 2015," Taylor Ahlstrom. Source: National Priorities Project interactive tax calculator, https://www.nationalpriorities.org/interactive-data/taxday/average/2021/receipt/.

★★★

80   Julia Kirschenbaum, Michael Li, "Gerrymandering Explained," Brennan Center for Justice, last modified August 12, 2021, https://www.brennancenter.org/our-work/research-reports/gerrymandering-explained.

81   Martin Gilens, Benjamin I. Page, "Testing Theories of American Politics: Elites, Interest Groups, and Average Citizens," *Perspectives on Politics* 12, no.3 (September 2014): 564–581, https://doi.org/10.1017/S1537592714001595.

82   Dan Kopf, "Why TV shows like "The Simpsons" and "Grey's Anatomy" keep getting shorter," Quartz, September 13, 2020, https://qz.com/1897841/tv-ads-are-making-shows-shorter.

83   B Corporation Homepage, accessed April 20, 2023, https://www.bcorporation.net/en-us.

84   "Total lobbing spending in the United States from 1998 to 2022," Statista, April 5, 2023, https://www.statista.com/statistics/257337/total-lobbying-spending-in-the-us/.

85   "Revolving Door: Former Members of the 115th Congress," Accessed April 18, 2023, Open Secrets, https://www.opensecrets.org/revolving/departing.php?cong=115.

86   Jennifer Rankin, "Is this the beginning of the end for fast fashion? The EU is moving to clean up the dirty industry," *The Guardian*, March 30, 2022, https://www.theguardian.com/environment/2022/mar/30/eu-wants-to-force-fashion-firms-to-make-clothes-more-durable-and-recyclable.

87   Clint Rainey, "The age of 'greedflation' is here: See how obscene CEO-to-worker pay ratios are right now," July 18, 2022,   https://www.fastcompany.com/90770163/the-age-of-greedflation-is-here-see-how-obscene-ceo-to-worker-pay-ratios-are-right-now.

88   Jeanne Sahadi, "What your 2015 income tax dollars paid for," April 18, 2016, CNN Money, https://money.cnn.com/2016/04/18/pf/taxes/how-are-tax-dollars-spent/.

89   Robert Frank, "The wealthiest 10% of Americans own a record 89% of all U.S. stocks," CNBC, Updated October 18, 2021, https://www.cnbc.com/2021/10/18/the-wealthiest-10percent-of-americans-own-a-record-89percent-of-all-us-stocks.html.

90   Jason Hickel, "To deal with climate change we need a new financial system," *The Guardian*, November 5, 2016, https://www.theguardian.com/global-development-professionals-network/2016/nov/05/how-a-new-money-system-could-help-stop-climate-change.

91   Nicole Coscolluela, Megan Cullen, Jason Norris, et al., "Beyond Black and White: Visualizing Charlotte's Busing Story, North Carolina State University, Accessed April 20, 2023, https://beyondblackandwhite.github.io/.

The Cultural Shift We Need

Where are Maynard Keynes' grandkids?

Epigraph: Henry David Thoreau, *Walden* (Boston: Ticknor and Fields, 1854).

Image 11.1: Eva Basilion (@EBasilion), Twitter, November 26, 2022, https://twitter.com/EBasilion/status/1596559385580670976.

★★★

92   Margie Omero and Ariel Kaminer, "'Listen to Us.' What These 12 Kids Want Adults to Know," *New York Times*, March 21, 2023, https://www.nytimes.com/interactive/2023/03/21/opinion/teen-youth-focus-group.html.

93   Nicholas D. Kristof, Sheryl WuDunn, *Tightrope: Americans Reaching for Hope* (New York: Knopf, 2020).

94   John Stuart Mill, *Principles of Political Economy with Some of Their Applications to Social Philosophy* (London: John W. Parker, West Strand, 1848).

95   John Creamer, Emily A. Shrider, Kalee Burns, et al., "Poverty in the United States: 2021," U.S. Census Bureau, September 13, 2022, https://www.census.gov/library/publications/2022/demo/p60-277.html.

96   Jamie Dunaway-Seale, "U.S. Rent Prices Are Rising 4x Faster Than Income (2022 Data)," Real Estate Watch, May 16, 2022, https://www.realestatewitch.com/rent-to-income-ratio-2022/.

97   Peter Whybrow, *American Mania: When More Is Not Enough* (W.W. Norton & Company, 2005).

98   Adam Smith, *A Theory of Moral Sentiments*, printed for Andrew Millar, in the Strand; and Alexander Kincaid and J. Bell, in Edinburgh, 1759.

Fetch the Bolt Cutters

Epigraph: Langston Hughes, "Let America Be America Again," *A New Song,* Knopf, 1938.

Image 12.1: "What would you say you do here?" meme adapted by Taylor Ahlstrom. Source material: *Office Space, directed by Mike Judge, featuring John C. McGinley, Paul Wilson, 20ᵗʰ Century Studios, 1999.*

Figure 12.2: "The Productivity-Pay Gap," adapted by Taylor Ahlstrom. Source: EPI analysis of Bureau of Labor Statistics and Bureau of Economic Analysis data. Economic Policy Institute, https://www.epi.org/publication/charting-wage-stagnation/.

★★★

99   "Our Living Standards Framework," New Zealand Treasury, Updated April 12, 2022, https://www.treasury.govt.nz/information-and-services/nz-economy/higher-living-standards/our-living-standards-framework.

100   John Kenneth Galbraith, *The Affluent Society* (Houghton Mifflin, 1958).

101   Sky Ariella, "36 Automotive Industry Statistics [2023]," March 15, 2023, Zippia, https://www.zippia.com/advice/automotive-industry-statistics/.

102   "Few Rewards: An agenda to give America's working poor a raise," Oxfam, Accessed May 24, 2023, https://www.oxfamamerica.org/explore/countries/united-states/poverty-in-the-us/low-wage-map/.

103 Sharon Block, "Supreme Court rules against union over strike liability," SCOTUSBlog, June 2, 2023, https://www.scotusblog.com/2023/06/supreme-court-rules-against-union-over-strike-liability/.

104 Jason Hickel, *Less Is More: How Degrowth Will Save the World* (Windmill Books, 2020).

By Design or Disaster

Epigraph: Nicholas Georgescu-Roegen, "Energy and Economic Myths," *Southern Economic Journal*, 41, no. 3 (January 1975): 347-381, https://doi.org/10.2307/1056148.

★★★

105 "BP Statistical Review of World Energy 2022, 71st edition," BP, June 2022, https://www.bp.com/en/global/corporate/energy-economics/statistical-review-of-world-energy.html.

106 Tim Gore, "Confronting carbon inequality: Putting climate justice at the heart of the COVID-19 recovery," Oxfam, September 21, 2020, https://www.oxfam.org/en/research/confronting-carbon-inequality.

Enough Is Enough

Epigraph: Henry David Thoreau, *Walden* (Boston: Ticknor and Fields, 1854).

★★★

107 William Morris, "Useful Work Versus Useless Toil," *The Collected Works of William Morris* (Cambridge University Press, 2013).

108 "Consumer expenditures report 2019," U.S. Bureau of Labor Statistics, December 2020, https://www.bls.gov/opub/reports/consumer-expenditures/2019/home.htm.

109 Philip Brickman, Dan Coates, and Ronnie Janoff-Bulman, "Lottery winners and accident victims: Is happiness relative?" *Journal of Personality and Social Psychology* 36, no. 8, (1978): 917–927, https://psycnet.apa.org/record/1980-01001-001.

110 Taylor Ahlstrom, *Authentic AF* (Emdashery Books, 2022).

111 James Suzman, *Affluence without Abundance: What We Can Learn from the World's Most Successful Civilisation* (Bloomsbury USA, 2017).

Innovating Our Way Out

Epigraph: Aldo Leopold, *A Sand County Almanac: And Sketches Here and There* (Oxford University Press, 1949).

★★★

112 "Global Land Outlook 2," United Nations Convention to Combat Desertification, April 27, 2022, https://www.unccd.int/resources/global-land-outlook/global-land-outlook-2nd-edition.

113 Peter Cosier, Tim Flannery, Ronnie Harding, et al., "Optimizing Carbon in the Australian Landscape," Wentworth Group of Concerned Scientists, October 2009, Accessed June 21, 2023, http://wentworthgroup.org/wp-content/uploads/2013/10/Optimising-Carbon-in-the-Australian-Landscape.pdf

Fuck it, dude. Let's go bowling.

Epigraph: Martin Luther King, Jr., speech in St. Louis, March 22, 1964.

★★★

114 Jerry Z. Muller, *Adam Smith in His Time and Ours: Designing the Decent Society* (Princeton University Press, 1995).

115 Julianne Holt-Lunstad, Mark Baker, Timothy B. Smith, et al., "Loneliness and Social Isolation as Risk Factors for Mortality: A Meta-Analytic Review." *Perspectives on Psychological Science* 10, no. 2 (March 2015): 227–237, https://doi.org/10.1177/1745691614568352.

Material Girl in a Material World

Epigraph: Fredric Jameson, "*Future City,*" *New Left Review* 21, (May 1, 2003): 65.

★★★

116 Matthias Schmelzer, Aaron Vansintjan, Andrea Vetter, *The Future is Degrowth: A Guide to a World Beyond Capitalism* (Verso, 2022).

117 Debbie Bookchin, "Radical Municipalism: The Future We Deserve," *Roar*, 6 (2017), https://roarmag.org/magazine/debbie-bookchin-municipalism-rebel-cities/

118 Michael Crichton, "Environmentalism Is a Religion," remarks to the Commonwealth Club, San Francisco, September 15, 2003, https://www.hawaiifreepress.com/Articles-Main/ID/2818/Crichton-Environmentalism-is-a-religion.

Where Do We Start?

Epigraph: George Washington Burnap, *The Sphere and Duties of a Woman: A Course of Lectures* (Baltimore: J. Murphy, 1854).

Hedonism is a Reasonable Response

Epigraph: Samuel Beckett, *The Unnamable*, English translation (Grove Press, 1958).

Image 13.1: Mike Primavera (@primawesome), Twitter, January 11, 2022, https://twitter.com/primawesome/status/1480959114289291266.

Hard Times Call for Furious Dancing

Epigraph: E.B. White, "Notes and Comments by Author," Israel Shenker, *New York Times*, July 11, 1969, https://archive.nytimes.com/www.nytimes.com/books/97/08/03/lifetimes/white-notes.html.

Index

About the Author

Taylor Ahlstrom is a writer, wanderer, and part-time hedonist. When she's not busy trying to save the world, you can find her savoring it with her husband and friends in Madrid, Spain. *This Isn't Fine* is her second book, and she's already started writing number three.

★★★

Head to taylorahlstrom.com/this-isnt-fine to access full-color, high resolution images from the book, and to download your "End of the World as We Know It" BINGO card set!

taylorahlstrom.com
@nomadiam